America's Classic
FARM TRACTORS

FORMERLY
FARM TRACTORS: A LIVING HISTORY

RANDY LEFFINGWELL

MBI Publishing Company

THIS BOOK IS DEDICATED TO DR. REYNOLD M. WIK, OAKLAND, CALIFORNIA

Dr. Wik is Professor Emeritas of American History at Mills College, Oakland, California, from which he retired in 1975. He returned then to his undergraduate alma mater and began a new career as Professor of American History at Sioux Falls College, Sioux Falls, South Dakota. He retired from there in 1985. Dr. Wik earned his PhD in American History from the University of Minnesota in 1949. His thesis became the first of three books published during his career, Steam Power on the American Farm, printed in 1953. He has also authored Henry Ford & Grass-roots American, published in 1972 and Benjamin Holt & Caterpillar Tracks & Combines, published in 1984. In addition, he has written and had published more than 45 articles, making him one of the most significant and important historians of mechanized agriculture in North America.

This edition published in 1999 by MBI Publishing Company, 729 Prospect Avenue, PO Box 1, Osceola, WI 54020-0001 USA.

Farm Tractors: A Living History first published in 1995 and *Classic Farm Tractors* first published in 1996 by MBI Publishing Comapny, 729 Prospect Avenue, PO Box 1, Osceola, WI 54020-0001 USA.

© Randy Leffingwell, 1995, 1996, 1999

All rights reserved. With the exception of quoting brief passages for the purposes of review no part of this publication may be reproduced without prior written permission from the Publisher.

The information in this book is true and complete to the best of our knowledge. All recommendations are made without any guarantee on the part of the author or Publisher, who also disclaim any liability incurred in connection with the use of this data or specific details.

We recognize that some words, model names and designations, for example, mentioned herein are the property of the trademark holder. We use them for identification purposes only. This is not an official publication.

MBI Publishing Company books are also available at discounts in bulk quantity for industrial or sales-promotional use. For details write to Special Sales Manager at Motorbooks International Wholesalers & Distributors, 729 Prospect Avenue, PO Box 1, Osceola, WI 54020-0001 USA.

Library of Congress Cataloging-in-Publication Data Available

ISBN 0-7603-0822-5

On the front cover: Oliver built some of the most stylish tractors ever to plow a field. The Model 88 Standard was introduced in late 1947 (shown here) remaining in production until 1954.

On the back cover: Top: Case's Model VAC was introduced in 1942 and offered into the early 1950s. Bottom: Detail showing the farmer's view of a 1949 "wide-front" Model VAC.

Printed in Hong Kong

Christmas 1999 - To Don from Jim & Anita Leigh Meyerings

Contents

FARM TRACTORS: A LIVING HISTORY

Acknowledgments 6

Introduction
Twentieth Century Entertainment 8

Chapter One
Horse Farming Draws to a Close 10

Chapter Two
Dawn of Internal Combustion 26

Chapter Three
The Nebraska Tractor Tests 60

Chapter Four
Development of the Diesel 78

Chapter Five
A Rubber Revelation 112

Chapter Six
Fuel Evolution 140

Chapter Seven
Industrial Design Comes to the Tractor 150

Chapter Eight
Ford, Ferguson, and the Three-Point Hitch 184

Chapter Nine
A Brief History of Tractor Innovation 202

Index 431

Contents

CLASSIC FARM TRACTORS

	Acknowledgments	245
	Introduction	247
Chapter 1	Allis-Chalmers	255
Chapter 2	Case	271
Chapter 3	Caterpillar	293
Chapter 4	Deere	311
Chapter 5	Ford	331
Chapter 6	International Harvester	347
Chapter 7	Massey	371
Chapter 8	White	379
Chapter 9	Orphans	403
	Index	432

Acknowledgments

I AM VERY GRATEFUL TO THE COUNTLESS TRACTOR COLLECTORS, historians and enthusiasts in the United States and Canada for their boundless cooperation. In particular, I owe sincere thanks to the following individuals for their generous help in producing this book.

Mike Altman, Wesley, Iowa; Frank and Evelyn Bettencourt, Vernalis, California; the late Donald "Tiny" and Alice Blom, Manilla, Iowa; Paul Brecheisen, Helena, Ohio; Jerry and Alma Clark, Ceres, California; Rob W. Collins, Laoma, Wisconsin; Don and Patty Dougherty, Colfax, California; Paul, Ray and Willard Ehlinger, Wabeno, Wisconsin; Dwight and Katy Emstrom, Galesburg, Illinois; Ray and Dorothy Errett, Harlan, Iowa; Palmer and Harriett Fossum, Northfield, Minnesota; Edith Heidrick, Woodland, California; Walter and Lois Keller, Kaukauna, Wisconsin; and Bruce and Judy Keller, Brillion, Wisconsin; Paul Kirstein, Clarion, Iowa; Lester Larsen, Lincoln, Nebraska; Lester, Kenny and Harland Layher, Wood River, Nebraska; Larry and Melanie Maasdam, Clarion, Iowa; Clyde and Jeanette McCollough, Vail, Iowa; Jerry Mez, Avoca, Iowa; Roger Mohr, Vail, Iowa; Rodney Ott, Hilbert, Wisconsin; Gene Pionek, Wabeno, Wisconsin; Bob and Mary Pollock, Denison, Iowa; Raymond and Dorothy Pollock, Vail, Iowa; Henry Roskilly, Tavistock, Devon, England; Carlton Sather, Northfield, Minnesota; Randy and Monica Sawyers, Shelby, Iowa; Eugene F. Schmidt, Bluffton, Ohio; Raynard and Ruth Schmidt, Vail, Iowa; Wes, Bonnie and Scott Stoelk, Westside, Iowa; and Kermit Wilke, Wilcox, Nebraska.

In addition, my thanks goes, once again, to Lorry Dunning, Historical Consultant, Davis, California, for his tireless and speedy help.

I am grateful also to the Department of Special Collections, University of California, Davis, Library, and to its director, John Skarstad, for his continued support and assistance.

My sincere thanks go to Ms. Kim Kapin, General Manager, and Jim Yates, Director of Marketing of A&I Color Labs, Hollywood, California, for their constant critical care and watchful handling of all of my Kodachrome.

Finally, I am extremely indebted to Larry Armstrong, Director of Photography, Terry Schwadron, Assistant Managing Editor and Shelby Coffey III, Editor, *Los Angeles Times* for granting me the leave of absence during which time I worked on this project.

Randy Leffingwell
Los Angeles, California

Introduction

Twentieth Century Entertainment

In a farm field one mile south and just off as far west of Exit 40 on Interstate 80 near Avoca in western Iowa, a small group of people waited impatiently for dark. It was just after Labor Day, a time of year when the sun settles to the ground at almost 275° by the compass and takes until nearly 8:50 p.m. to get there.

A stiff breeze had come up about 6 p.m., and some of the people wondered what effect it might have on the evening's activities. The wind blew hard out of the south at a steady 15–20mph. The sky filled with flat, gray-bottomed clouds that looked as though they had been scraped along the earth. The clouds were soiled and filed flat on the bottom, with billowing tops inflating upwards. Out at the horizon, though, the sky remained clear and blue. A few people looked to the north, concerned about the consequences of what was to come after dark.

The corn would be okay, they agreed, because it was still moist. The nearest house was a half mile beyond the corn, safely out of harm's way. The grass underneath their feet was dry, but rain during the week left everything too green to burn.

A pickup truck arrived loaded with nine Danish modern brown-stained wood chairs. The sagging foam cushions were covered in seventies knobby tweed. Folding metal chairs filled in the corners of the truck bed. Unloading it was like trying to remove one coat hanger from a bundle. Half the load came up at one tug; men joined in to untangle and unload. The chairs were set up in a diagonal line, slightly to the north and about 20ft east of the 1913 Russell 20hp steam traction engine that stood gently coughing at idle. A thin string of gray smoke rose from its stack. Abruptly cut off by the wind, it whipped to the north.

In the increasing gloom, women gathered children and flashlights. It would be too dark to see the way to cars or campers when they left. In the growing shadows forms moved, burdened with thermos bottles and six-packs of sodas or beer. The crowd grew in modest numbers. They were mostly friends of those participating in the threshing bee that had gone on all day, invited guests of the hosts of the evening.

"Will they still do it? Even with this wind?"

In answer, the Russell cracked the dusk with a long blast of its steam whistle. Randy Sawyers rolled his wrist on and off the cord, playing the single-note whistle like a Wurlitzer theater organ. Still, it was not yet fully dark.

There was activity around the Russell. Faceless figures moved at the front and rear. A hopper of kindling, wood shavings, and sawdust had been moved in the afternoon to a position near the back of the steam engine. Seventy feet forward was a low windmill, anchored to the ground with circus tent stakes. Its four paddles, each 2ft square, were meant to fight the wind even on calm days. It was loosely connected by a canvas belt to a large pulley wheel on the Russell. These devices were used to give a measure of an engine's performance. Knowing the diameters of the tractor and windmill pulleys and using a hand-held tachometer pressed against the windmill axle hub to count the revolutions, it was possible through a few simple calculations to determine "belt pulley horsepower."

But tonight, no one was counting.

Raleigh Woltmann reached up a gloved hand. He opened the front access door of the steamer's long, horizontal firebox. A volcanic glow lit his face. It was fully dark at last. The stars and moon were obscured by the cover blown in from Kansas.

Flashlights went on. Pools of light jiggled across the field toward cars. Doors opened and parents yelled to children to come put on sweaters and jackets to guard against a chilly south wind that made the night cool for early September. Feet shuffled through the grass to and from the Danish modern seating group. People returned, bundled in warmer clothes.

Sawyers opened the throttle on the Russell and, downwind, the windmill began to turn. It made white noise and tried to force back the low-pressure front moving in overhead. Woltmann swung the door open, and 5ft above him a few sparks shot out of the broad, trumpet-belled stack. The glowing cinders went up and quickly arced to the north. Sawyers opened more throttle, and a few more sparks shot like tracer rounds into the night to chase the others.

"Don't get excited folks," Woltmann yelled over the roar of the winds of God and Russell. "That's just the ashes inside the firebox. We haven't begun yet."

Sawyers opened the throttle to near full power. The chuffing sounded like the soundtrack from a railroad chase movie. The Russell, glowing from the front, turned orange at the rear as Sawyers checked its fire. The large machine at full speed, outlined in its own glow, ran evenly; it rested rock steady as though cemented to the ground.

Then Sawyers began. He shoveled a scoop of wood shavings into the fire door and Woltmann's face flashed white. A vesuvian molten stream of fire roared past the boiler tubes and curved like a liquid up into the stack. A column of yellow-orange streamers wound madly up into the sky, caught the wind, and rolled north toward the interstate a mile away.

Another scoop and another and another loaded up the firebox with flash fluff. A deep moan slipped out the front door past Woltmann. The light inside the Russell's round belly glared white. It cast Woltmann's shadow on the ground downhill toward the windmill and beyond, all the way to the corn.

Errant tracers, fallen out of the jet stream, glided into the turmoil of the windmill and were spun back into the air or down into the ground, or else off on wildly skewed trails of their own making.

The crowd in the Danish modern gallery longed out loud for popcorn. And they wondered more softly if city-folk got this much enjoyment out of Fourth of July firecrackers. Scoop after scoop of sawdust and wood chips went into the white heat and streamed out the black stack as the tractor labored on the belt, turning the windmill blades into blurs in the dark.

Interstate travelers pass this spot a mile to the north and are usually greeted with a six-mile stretch of seamless darkness to the south. Tonight, traffic in both directions was slowed by the spectacle. The fiery liquid orange spout was hurled 50ft up by the Russell's draft and roar and carried 100 yards downwind, like fireworks trailing off into the acrid smoke.

For another twenty minutes Sawyers and Woltmann played the Russell. One heaved shovels full of wood shards into the fire. The other controlled the draft and that sound, that deep bass moan of the draft, playing the front door like the wind instrument it was. For twenty minutes, traffic a mile away slowed as it crossed America and wondered if what it saw was real or just a product of road weariness.

When the gusts were interrupted by a lull, sparks soared like aerial skyrockets and fell back on themselves like the best of fireworks. Then the wind would pick up, and the orange geyser would sweep away in a tall flaming wall, heading toward the amazement of confused viewers, families, and truckers pulled to the shoulders along Interstate 80.

Then it was over. The last few weak fireflies turned tail and flamed off. Or they got tripped up and spun around by the windmill. A cheer and a round of applause went up. It was almost as loud and boisterous as any New York City or San Francisco field of spectators rising in praise of the Independence Day fireworks display.

Flashlights went on, and families drifted back to their cars or over to their campers. The chairs went back into the pickup bed. Sawyers and Woltmann moved around, shutting down the Russell for the night. The hellish fire had emptied the firebox, but the head of steam in the boiler would take longer to calm down.

Along the interstate, traffic began to move again. A huge semitrailer, visible even from the hillside, flashed its orange trailer lights rapidly.

"What do you suppose they all thought this was?" someone asked in the dark.

A long whoosh of steam bled off the boiler. Sawyers ripped two short and one long blast through the whistle into the night. It was the signal that steam work had finished.

"A field fire," a voice said matter-of-factly, hefting a Danish modern back into the pickup.

"Naw, fire wasn't movin'," said a voice farther away. "Prob'ly more likely a barn fire!" the voice offered.

"Too small," a woman replied. "And barns don't shoot up sparks from only one place."

"A volcano?" a younger voice spoke. Older voices laughed. "Well, it could be," the younger voice defended itself. "Mount St. Helens shot flames in the air."

The memory of the orange column sweeping up out of the black night and curving to the north came back clearly. So did the deep heaving groan as the wind roared through the firebox and was sucked up the stack.

"It was just a sparks show," Randy Sawyers said, his voice aimed toward the younger one. "That's all it was."

In the deep blustery dark, the younger voice tried on other options. "Mount Saint Russell?" A woman nearby chuckled.

"Come on, silly. It's past your bedtime."

"I've got it," the young voice trailed off into the wind. "Krakatoa, Western Iowa…"

One advantage of historical study and antique restoration is that students and practitioners often can find pleasure and amusement in the objects and tasks that were once used purely for work. It is hard to imagine that the rancher in Alberta or farmer in Kansas who was fortunate enough to own a 20hp steam traction engine in the early part of this century would have fired it up for evening entertainment. The risks that those sparks—wind-driven up from Oklahoma or down from the Arctic Circle—would destroy a few thousand acres preclude taking the chance, even if the owner was disposed to the light-heartedness it evokes. The development of the farm tractor has progressed so far, however, that a "sparks show" for friends and neighbors is one of the likely tasks for a steamer. It is the same with horses. Farming with horses or mules at this end of the twentieth century is something done for religious reasons on farms that function to provide family livelihood. Or else they work on farms that serve as a diversion and a hobby for their owners. When horses were no longer beasts of agricultural burden or urban commerce, riding them became a pleasure activity. The techniques of breeding draft horses for strength were modified to the big business of breeding race horses for speed.

The history of agriculture progressed for thousands of years with oxen, horses, and mules before steam was tamed. A few decades later, gasoline replaced steam. Barely two more decades passed before liquefied petroleum gases were brought under control, and at nearly the same time, Dr. Rudolf Diesel's engines were introduced. Within four more decades, diesel power's strength and reliability made every other fuel source obsolete.

What uses will Ford's articulated four-wheel-drive tractors or Caterpillar's rubber-tracked Challengers serve in another half century? Will they do more than fill collectors' sheds and museums? What will have replaced them in daily use? It is entertaining to speculate as to what nostalgic purposes their replacements will be put in the year 2100.

Chapter One

Horse Farming Draws to a Close

TAKING POWER OFF THE HOOF

OTTO VON GUERICKE LAUNCHED THE world into self-propelled motion roughly 350 years ago in Germany. While it would take nearly another 200 years for machines derived from his discoveries to move from their own power, it was the results of his work that encouraged the possibilities.

Von Guericke was aware of the axiom that nature abhors a vacuum. He theorized that there was no air in outer space, because if there was, its friction would have slowed the travel of planets and stars. To satisfy his curiosity, he set out to create a vacuum. The fifty-year-old philosopher took a hollow brass sphere and something like a bicycle pump to suck the air *out* of the sphere. He watched his sphere, and even though he imagined what might happen, he was shocked when it did. After most of the air had been pumped out, the sphere crumpled in on itself.

Von Guericke had discovered atmospheric pressure, an element that would prove crucial to the functioning of internal-combustion engines. He continued experimenting with larger pumps and spheres. His pump consisted of a piston and connecting rod inside a cylinder with valves. At one end of the closed pump was the sphere; at the other, the connecting rod protruded through a sealed cap to a handle. Using this kind of simple "engine," he engaged his own arm muscles as the fuel for the driving force to pull out the pump handle. Releasing the handle at the top of its pull away from the sphere, he watched as the connecting shaft was pulled back into the cylinder, by itself. This result was energy without human, wind, or water power. Of course, the vacuum—the low air pressure—created inside the sphere was simply pulling back its own air from inside the pump. The piston, pulled up by von Guericke, was pulled back by the vacuum.

▲*The Phoenix Log Hauler tender held about 315gal with another 400–450gal in the tank. The steam operated two 6.5x8in compound cylinders on each side, producing a total of 100hp. This power was delivered equally by a single throttle-valve to all four cylinders. Tracks were iron, with twenty-nine pads per side, moving on Cletrac-like track-rollers. Track gauge was 65in.*

▶*Paul Ehlinger watches the stack blow above him. This 1909 Phoenix Log Hauler runs summer demonstrations in Wabeno, Wisconsin, where this and two other identical machines were owned by G.W. Jones Lumber. This machine, #79, was the lumber mill yard tractor. The only steering was accomplished by this wheel; there were no track brakes or clutches; a main-drive clutch stopped or engaged forward or rearward motion.*

A young Dutch physicist, Christiaan Huygens, lived in Paris at this time, around 1650. Huygens and his younger French assistant, Denis Papin, were both born and raised during the Thirty Years War, a conflict that introduced western civilization to gunpowder. Huygens was a pacifist with an imagination. He had invented the pendulum clock and perfected the telescope.

Envisioning French King Louis XIV's cannons as a kind of open-ended air pump of the type von Guericke had experimented with, Huygens imagined the cannonball as a surrogate piston. He reasoned that if gunpowder was placed below the cannonball in a sealed cannon tube and the gunpowder was ignited, then at the top of the sealed tube, the air compressed by the rising ball—Louis XIV's piston—would force it back down. Remarkably, neither Huygens nor Papin were killed during their experiments. The efforts proved Huygens correct. For a grander version of his experiment, his piston was fixed to a rod through the cannon's sealed end, onto which a rope was then attached. In a demonstration for the French government, Huygens lit the fuse. The explosion inside the tube raised the piston. This moved the rope and lifted a platform loaded with men. It was an internal-combustion engine that produced heat. It had now put that heat energy to work.

But fresh gunpowder could not be continuously injected into a hot cylinder without disastrous results. Papin realized that steam made in the cylinder could force the piston away as well, compressing the air at the other end. When the steam cooled, the piston returned. But Papin's imagination didn't extend far enough. He did not conceive of an outside source of steam that was fed continuously into the cylinder and that would be bled off on the piston's return. Even when the idea came to him, Papin

was plagued by pipes that leaked, seals that failed against the pressure, and seventeenth century technology that was unable to cast a perfectly round cylinder.

The lack of perfect cylinders plagued Thomas Newcomen in Devonshire, England, as well. For something like a decade, from around 1698 probably until sometime in 1708, he and his assistant, John Calley, labored to produce a pump to remove water flooding the coal mines. A Londoner, Thomas Savery, a Newcomen contemporary, had done work with steam that he heated and condensed in two separate chambers to produce a nearly continuous suction pressure in hopes of withdrawing water from great depths out of the mines. Pipes fitted with one-way valves extended down into the mine water and up out the top to a run-off. But Savery's system required great boiler pressure. His soldered seams occasionally melted from the heat generated by the boiler fire, and the steam and condensed water had to be vented by hand valves operated by an attendant usually stationed in the middle of the mine-shaft. Savery's pumps were estimated to be successful at sucking up water as much as 150ft. Little evidence exists, however, that they did better than 20ft. He gave up all efforts at mine pumps after 1705.

Newcomen may or may not have known of Savery's work. Records of Newcomen's early life do not exist. It is 160mi from Devon to London, and in the seventeenth century, news of the success or failures of inventions traveled slowly. It is more likely that the stories of Papin's work were known widely. Their coincidental developments were simply a case where several inventors, obsessed by the same need, set the same goals and arrived at similar conclusions.

Newcomen understood that the condensation of the steam in his cylinder drew the piston pump back as the pressure inside sought to equal the atmospheric pressure outside. He realized that relying on the outside air to cool his cylinder slowed his engine's work to an imperceptible pace. He first tried cooling the cylinder by pouring cold water over the outside walls. Then he tried to jacket the cylinder and control the cooling flow surrounding it. But this proved to be barely any improvement over air cooling. Then one day, one of Newcomen's soldered repairs in his cylinder gave way. A tiny hole allowed cold water to stream rapidly into the cylinder. This cooled the vapor quickly. The connecting rod had been attached to a balance beam and to a weight, duplicating on a smaller scale the work achieved by Huygens' cylinder. The accidental cold-water injection condensed Newcomen's steam so fast and so forcefully that the chain broke. The piston crushed the bot-

▼*The left front ski of the Phoenix Log Hauler emerges from the bleed-off of a full head of steam. The Phoenix Lumber Company of Eau Claire, Wisconsin, licensed the production of this example and more than sixty others from lumberman/ inventor Alvin Lombard of Maine. Lombard and Phoenix eventually produced about 215 of these machines, used across the northern United States and into Canada. Lombard later ended up in a furious patent dispute with Caterpillar inventor Ben Holt over crawler-tractor design.*

▲The coldest job in North America was reputed to be that of tillerman on the Lombard or Phoenix in a Maine, Ontario, or Wisconsin winter. The Log Hauler was capable of hauling out sleighs loaded with a dozen or more 15–16ft tree sections at speeds of 6–8mph. But all the heat was behind the tillerman, who sat out in the elements. In Maine, there were stories of putting an outhouse on the platform around the tillerman and the assistant.

tom of the cylinder and the lid of the boiler as well.

Thomas Newcomen, an ironmonger by trade, a metal-craftsman in fact, had succeeded in making a rapid-acting steam engine. However, it had acted only for one cycle. Hand-operated valves had been fitted to the top of the boiler to allow fresh cold water in to replace the quantity lost to steam. It was now Newcomen's chore to conjure a method of making these valves fast-acting and, indeed, self-activating, if this engine was ever to be capable of continuous work.

John Farey, an architect and engineer in the early 1700s, witnessed a production Newcomen engine, and his "Treatise On the Steam Engine," is quoted in R. L. T. Rolt's 1963 book, *Thomas Newcomen: The Prehistory of the Steam Engine*. "At first the valves were opened and shut by hand," Farey observed, "and required the most exact and unremitting care of the attendant to perform those operations at the precise moment; the least neglect or inadvertence might be ruinous to the machine, by beating out the bottom of the cylinder, or allowing the piston to be drawn wholly out of it."

Dr. William Stukeley, a London physician, visited a Newcomen pump installation in Whitehaven, England, in 1725 and described its workings. Rolt quoted from Stukeley's notes. "It creates a vacuum by first rarefying the air with hot steam, then condenses it suddenly by cold water," Stukeley wrote, "whense [sic] a piston is drawn up and down alternately, at one end of the beam: this actuates a pump at the other end, which, let down into the works, draws the water out: it makes about fourteen strokes a minute."

According to Rolt, an engineer named Henry Beighton invented a linkage that he had installed on a Newcomen-style pump in 1718. The linkage ran from the pump's balance beam to the valves and opened and shut each of them in time to the action of the pump and the steam boiler. It was, in effect, one of the earliest valve rocker-arms, establishing an engine mechanism that has been in use ever since. With Beighton's improvement, Newcomen's steam pump became a reliable steam engine, ready for the next step in its development. This would translate the vertical shaft movement to rotary motion.

Because the stroke length of the Newcomen engine varied with the amount of steam admitted to or expelled from the cylinder, Newcomen's contemporaries failed to consider that a crankshaft and flywheel might serve to even out the stroke and average out the production of power. Instead, they felt that its irregular stroke prohibited the engine from becoming a rotating power source. It was not until 1763 that Joseph Oxley of Northumberland patented a ratchet device that prevented the crankshaft from turning backwards if the steam pressure dipped for any period of time. Yet, Oxley's ratchet device was flawed, as were those that followed during the next several years.

James Watt, a Scottish engineer and scientist, made the improvements that turned Newcomen's beam engine into a practical machine that would fit many more applications. His first idea, patented in 1769, separated the condenser function from the piston cylinder. This meant that the cylinder could remain near steam heat all the time. The condenser, shot with cold water, sucked the steam out of the cylinder, pulling the piston down

◂The Log Hauler stands 124in high, 73in wide, and it's 228in long overall. It was used by G.W. Jones Lumber Co. in Wabeno until 1929 and then sold to the City of Wabeno, Wisconsin, in 1935. It required a steam engineer, fireman, tillerman, and assistant to operate it. These days, volunteers such as certified steam engineer Rob Collins of Laona, Wisconsin, and tillerman Paul Ehlinger make do with hands full.

to the bottom of the cylinder. Then steam could be immediately injected into the hot cylinder chamber to raise the piston again. This increased the number of cycles possible per minute and decreased substantially the amount of fuel burned in heating steam for the cylinder.

On a suggestion from his assistant, William Murdock, Watt attached his piston to a crankshaft in order to convert the reciprocal piston movement into rotation. But an ex-employee stole that idea and patented it first, forcing Watt to adapt. He adopted a sun-and-planet gear system and used that until 1794, when the crankshaft patent entered public domain.

With both the sun-and-planet gear system (where the connecting rod ended in a geared wheel that rotated around a geared center shaft) and the crankshaft, it was possible to add a flywheel. This evened out the action of the piston in the cylinder, making it more useful and manageable for other tasks. Perfection of the duplex engine—essentially two pistons working within the same cylinder, one always being in compression stroke—followed quickly.

Probably the first attempt at putting to work the inventions

◂The Log Hauler, patented May 21, 1901, operated as a curious mix of steam traction engine and steam locomotive. It had no compressor to operate air brakes as on a locomotive. Neither did it have a clutch as found on some steam traction engines. Friction on ice or snow from the load—sometimes as much as 100,000 board feet on as many as twenty-five 12ft-wide sleighs behind it—stopped it effectively. For the demonstrations each summer, the front skis are replaced with steel wheels.

▲Daniel Best's big 110hp steamers were available either as 28ft long freight haulers or like this 22ft long agricultural version. The shorter, more maneuverable agricultural models operated on round drive wheels 8ft in diameter and 5ft wide. Wood or metal extensions some- times added another 12ft to wheel width on each side to keep the eleven-ton tractors from sinking into soft soil. This 1904 steamer is owned by Mrs. Edith Heidrick of Woodland, California.

of Newcomen and Watt appeared in railroad locomotives. The first self-propelled road vehicle was Nicolas Cugnot's Steam Road Wagon, completed in 1769. But the first known agricultural application of a steam traction engine—that is, an engine that could haul itself and possibly other vehicles as its normal function—came from an English locomotive builder, E. B. Wilson & Co., at the Railway Foundry in Leeds. Known as the "Farmer's Engine," it was designed in 1849 by Wilson & Co.'s chief designer Robert Willis.

The Farmer's Engine was a two-cylinder simple engine, with a 6.25x10.0in bore and stroke. According to Ronald H. Clark in his 1960 book, *The Development of the English Traction Engine*, the Wilson Farmer's Engine worked for several weeks towing and operating a small threshing machine northeast of London. The horizontal coal boiler operated at 45psi yet produced barely 4.3hp. It burned 61lb of coal per hour and evaporated 41gal/hr. Yet with its two-speed gearing, it was capable of 12mph on level ground. It dispensed with horses completely. It was self-steering and self-propelled, the two pistons operating tractor-length connecting rods attached to a crankshaft and by gears attached to the rear driving axle.

From 1857 through 1862, John Smith, a farmer, engineer, and machinist in central England, produced half-a-dozen chain-driven traction engines. Smith mounted his boilers so they could be pivoted, raised, or lowered from the rear, to keep the boiler level while climbing or descending. At normal working pressure of

▲Melanie Maasdam, of Clarion, Iowa, works her two Belgians behind a Jenny Lind cultivator. On the left, Jane, is a 1,750lb eleven-year old mare, and on the right is Mary, an 1,800lb ten-year old mare. A rule of thumb in horse farming is to use the larger horse on the right.

▲Neighbor Paul Kirstein rides a John Deere cultivator behind Melanie Maasdam's Belgians. Kirstein grew up on an eighty-acre farm that used horses well into the early sixties. He got back into it as a hobby. In plowing or cultivating, he recalled, "It was peaceful. All you heard was the earth cracking and the horses' hooves clodding."

▲*Horses have the same appeal to Melanie Maasdam as they do to her friend, Paul Kirstein. Melanie and husband Larry refer to the horses as their Gentle Giants. Here she mows the roadside grass using a John Deere No. 3 horse-drawn mower.*

60psi, Smith's traction engines produced 10hp. Smith learned his engineering skills working for John Fowler & Company of Leeds.

Fowler produced railroad locomotives, and the configuration of Smith's and later of Fowler's own steam traction engines benefited from their experiences. Fowler's traction engine in 1868, designed by David Greig, used a single duplex 6.5x12in cylinder and even drove all four wheels. The Fowler, Smith, and Wilson engines were all "undermounted"; that is, the engines' cylinders and flywheels were below the boilers.

But many other makers placed their cylinders on top of the boiler, and self-propulsion was accomplished through chains from the "overmounted" crankshaft to the large rear drive wheels. A number of these "portable" engines still relied on horses for steering, but history suggests that the horses often did more than steer.

"The underlying idea of making these heavy engines self moving," Clark wrote, "was to spare the horses much killing labour. I have had it from several sources that many fine animals have been strained and later necessarily destroyed as a result of getting bogged portables out of awkward places."

At the same time as E. B. Wilson's Farmer's Engine in England in 1849, a Philadelphian, A. L. Archanbault, introduced his Forty-Niner series for 4, 10, and 30hp horse-drawn portables. In 1850, Gideon Morgan of Calhoun, Tennessee, received a patent for improvements to a track-type tractor wheel substitute. This suggests that a crawler or tracklayer-style steamer existed at least in design before that time.

The first steam traction crawler in England was designed by Richard Bach of Birmingham in 1854. At the rear of one of his 8hp portable engines, he used the "Boydell's Patent Endless Railway," a kind of articulated plank arrangement that James Boydell first patented in 1846. William Tuxford produced several varieties of vertical boiler-equipped Boydell crawlers, and Charles Burrell produced many models of overmounted, horizontal boilers, complete with front- and rear-planked wheels, the huge rears being enclosed in sheet metal like an orchard tractor from a century later. Both Tuxford and Burrell Boydell-style steam traction engines were shipped as far away as Egypt and Brazil.

▸*This was a morning when the operators could sleep in. Randy Sawyers of Council Bluffs, Iowa, brings the Russell out for demonstrations and for a threshing bee in Avoca, Iowa, each September. Threshing has become entertainment rather than work, so everything waits till after sunup to begin.*

The technology and ideas crossed the Atlantic with the machines. In his 1976 book, *Encyclopedia of American Steam Traction Engines*, C. H. Wendel chronicled the history. In 1854, Henry Ames in Oswego, New York, founded one of the first factories in North America to build portable steamers. A year later, Obed Hussey produced his first steam plow in Baltimore, Maryland, and the race was on. Blandy Steam Engine Works in Zanesville, Ohio, offered portable engines from 3hp to 35hp, ranging in price from $300 to $2,300. Wheeler & Melick Co. in Albany, New York, produced 6, 8 and 10hp portables, operating at up to 200psi in 1877. But by then, companies such as the American Engine Company in Jersey City, New Jersey, had their first traction engines in production for two years. Henry Ames produced his first one in 1885.

By 1900, there were at least sixty-eight companies that had produced and sold steam traction engines in North America (although many had already gone out of business by that time). Seventeen of them were in Ohio alone. Twelve more were located in New York while five were established in Ontario.

Engineering features and developments ranged as far and wide as manufacturer locations. As early as 1871, D. D. Williamson in New York City produced a vertical boiler traction engine that was mounted on hard rubber tires. The Scottish inventor, R. W. Thompson, had developed the inflatable pneumatic tire in the late 1860s. Williamson contracted with Grant's Locomotive Works across the Hudson River in New Jersey, and fifty rubber-tired steamers were built.

The Lansing Iron and Engine Works, founded in Lansing, Michigan, in 1876 and out of business by about 1898, produced the 12hp Lansing Double Traction engine, a four-wheel-drive overmounted steamer. A marvelous jumble of chains drove both the solid front axle (that pivoted inside its large sprocket to turn) and the rear wheels.

Sawyer & Massey in Hamilton, Ontario, became Canadian agents for the English Aveling & Porter steam traction engines in 1887. A change was soon made to the locomotive-type boiler that would burn wood, coal, or straw. Throughout the 1890s, Sawyer & Massey produced traction engines ranging from 13hp up to 35hp. Within two decades, the company was producing 25 to 76hp compound engines, machines powerful enough to break sod in Alberta or operate sawmills in British Columbia.

In Massillon, Ohio, three brothers formed C. M. Russell & Co. as a general carpentry business. They quickly expanded into producing railroad cars and threshing machines, and, by the mid-1880s, portable 6hp steam engines. Almost immediately, the Russells had devised geared transmissions and chain-and-roller steering and produced 6hp and 10hp steam traction engines.

Russell followed common practice as technology and development continued around the world. The company overmounted the engines, placing cylinders up front to lessen the distance the steam traveled. This positioned the crankshaft above the center of the boiler, providing the minimum distance for gears to get power to the rear drive wheels. Russell tractors used a friction clutch and were equipped with two forward speeds and a reverse gear. Power quickly increased, and before 1900, Russell offered tractors with up to 16hp. Russell used single-cylinder configurations even for the 30hp engines of the early 1920s. The company introduced its 20hp model in 1912. For purposes of rigidity, the engine cylinder, steam chest, slide housing, and half the crankcase box were cast in a single piece, mounted atop the boiler. This was necessary since the largest Russell produced 67hp off the crankshaft-driven belt pulley. By the time the company

quit manufacturing steam traction engines, power had increased to 150hp, used more for road freight hauling than agricultural purposes.

In California, Dan Best and Ben Holt had the market largely to themselves because of the high cost of transporting traction engines around the Rocky Mountains. Soil conditions in the central California valleys necessitated huge wheel extensions, some as much as 26ft on a side. These were required to keep the heavy 110hp Best and Holt steamers on top of the peat bogs or sandy soils that California farmers had come to value. The developments of James Boydell with his planked wheels interested both Best and Holt very much.

Work by a Maine farmer and lumber mill owner, Alvin Lombard, attracted Holt and Best as well. Lombard had invent-

▶*Belted up to the fan, the Russell could put out much more than its drawbar-rated 20hp. This model used an 8.25x12in simple engine. The universal boiler contained thirty-six of the 2.5in tubes; this was different from the standard models fitted with about fifty of the 2in diameter tubes. Belt horsepower was nearer to sixty-five.*

ed a locomotive-type engine for use during the summer for plowing, but more important, during the winter as a log hauler. Front wheels were replaced by wooden skis, and the Log Hauler's crawler tracks provided unmatched traction. In one trip out of the Maine woods, Lombard pulled nearly 100,000 board feet of logs. The Phoenix Lumber Company in Eau Claire, Wisconsin, learned of Lombard's machines and bought one, and then negotiated the rights to manufacture them. Phoenix's Log Haulers used neither under- nor overmounted engines but instead fitted two cylinders vertically on either side of the long boiler. Worse, steering was accomplished by a tillerman seated outside on a bench mounted in front of the boiler. It was surely the coldest job imaginable. The 100hp engines were capable of phenomenal work over eastern and midwestern snow-covered logging roads. Phoenix eventually sold 200 of the Log Haulers, all with external steering.

Ben Holt learned of the Log Hauler's crawler tracks and adopted its technology to create his Caterpillars. Holt eventually built eight of the steam-powered crawlers, although both he and Daniel Best already had produced hundreds of wheel-type traction engines.

Huge steam traction engines worked well enough on farms of average size in the eastern and midwestern United States. They served even better on the vast ranches of the western states and provinces of Canada, breaking prehistoric sod for first planting. But in England, land was more scarce, and farms were much smaller. Even maneuvering a steamer with eight or twelve plows behind it required huge spaces, room that was better served in cultivation.

As early as 1800, patents were issued in England for frames carrying winding drums and cables or ropes to draw plows or cultivators across fields. By 1856, John Fowler & Co. had patented a system of cable-operated plows driven by a single steam traction engine and double rollers anchored to the ground on the opposite side of the field. This was first conceived as a way to plow wet ground to bury drainpipes. Version after version were tried, including placing the steamer on one bank of the field and a wheeled, movable windlass on the opposite bank. But each move took time. The cables had to be slacked, the anchors released, the windlass moved forward, and the anchors reset. By 1865, John Fowler had perfected a system using two of his engines, each fitted with a horizontal winding drum below the boiler. These were placed on opposite sides of the field, and while one played out fourteen-gauge steel cable, the other wound it up. A two-way plow was used so the furrows fell the same direction, and the two engines simply inched their way up the field headlands. Fowler sold not only plows but also cultivators and subsoilers to work this system, enabling his engines to perform work for which the steam traction engines were otherwise impossibly unsuited because of their size and weight.

But even with the enhanced efficiency that dozens of talented engineers and inventors obtained through Watt's rotary movement and his other improvements—slide valves for steam intake and exhaust and pressure gauges—the steam engine was immensely impractical. At its best, it produced only 4 percent thermal efficiency. That is the percentage of useful work an engine performs in comparison to the total energy content of the fuel consumed. Furthermore, it took as much as two hours to get one up to operating pressure. And it was dangerous: Leaks in pipes scalded operators; improperly at-

tended boilers exploded. The weight of the engines broke rural bridges, stranding the machines in sometimes irretrievable positions. The smoke from their fireboxes choked the air.

Still, with no alternative, operators from the time of James Watt would endure another one hundred years of dissatisfaction and experimentation before an alternative would appear. And it required an unrelated discovery in Pennsylvania in the 1850s to ensure that its replacement would succeed.

▶▶ *The 6ft x 12ft canopy was a $50 option at the time these steamers were new. A small wood cab was offered as well, for about $150. Without water in the boiler or wood in the bins, the 20hp model weighed 20,800lb.*

◂ *The Russell Company was founded in Massillon, Ohio, in 1838 and was famous first for making railroad cars and steam shovels. Their first steam tractor, a 6hp self-propelled, self steering model, was introduced in 1887. Russell remained in business until 1927, producing steam traction engines as large as 150hp. This 1913 20hp model was restored and is owned by Randy Sawyers of Council Bluffs, Iowa.*

Chapter Two

Dawn of Internal Combustion

Taking the Next Step

THE OIL DISCOVERED UNDER THE EASTERN United States was first processed for lamp fuel in the mid-1800s. Ironically, as gas processing improved and steam power was fully domesticated another fifty years later, gas that flowed in the headlamps lit the roads at night for steam-powered automobiles. This was a true technological contradiction if ever there was one.

Physicists, chemists, inventors, tinkerers, and experimenters all understood heat engines well enough by 1850. In fact, they knew a century earlier that to replace steam required new thinking. The heat that expanded the air to force the piston to move the crank to spin the flywheel *must* be created inside the cylinder. Internal combustion would eliminate the boiler, the firebox, and the steam. If only they could bring the firebox inside the cylinder and create the steam pressure instantaneously, and repeatedly, then they would have a practical engine.

Processed coal produced gas that was piped into homes for illumination. This was not without its own problems, of course, but its potential for small engines was obvious. The great industrial revolution had begun to pay benefits. Steam-powered belt-driven lathes and tools, operated and patented by iron craftsmen like John Wilkinson of England, turned out not only exquisite decorative sword blades but also precise cylinders and true pipes. New processes improved the quality of metals, blending and hybridizing them for strength and longevity. In laboratories throughout Europe and North America, electric sparks became more controllable. And in book-lined libraries and classrooms, under the attentive gaze of enraptured students and assistants, mathematicians and theoreticians demonstrated a better understanding of just what happened when heat produced energy. The patent offices got very busy.

▲*Bore and stroke was 7.5x9.0in. The giant Bear produced thirty drawbar and fifty belt pulley horsepower at 650rpm. The engine used a gear-driven oil pump to pressure-lubricate not only the crankshaft and cylinder walls but also the connecting rods and wrist pins. This was the third of nine ever produced, manufactured in late 1911, and it is owned and was restored by E.F. Schmidt of Bluffton, Ohio.*

▶*It was tractors like these that prompted agricultural journalists to call for more compact tractors. Its front wheel was actually a pair of 12x42in steel drums. The Bear measured 22ft long overall and stood 93.5in to the top of the exhaust. It was 99.5in wide and weighed 10.5 tons. Its four-cylinder 1,480ci engine was capable of pulling 8–10 14in plows. There was nothing small about the Wallis Bear.*

In 1791, Englishman John Barber received a patent for his gas turbine engine. Historian Lyle Cummins wrote of Barber's work in his 1976 book, *Internal Fire*. The discoveries of Christiaan Huygens and Denis Papin suggested to Barber that other flammable substances besides gunpowder could be used to power his engine. If ignited in a closed cylinder, these materials could turn the resultant heat into work. Barber used coal gas, but unlike Watt's developments with rotary motion, Barber used the force of the exhausted combustion. He channeled it through a small opening. This was directed toward a turbine wheel that he geared to an output shaft. Cummins' thorough research failed to turn up any working examples of Barber's early turbine, but the historian offered an insight. From Barber's patent it was clear that he had already imagined that his managed exhaust stream had enough force to be used as its own form of propulsion. This hinted at, as Cummins wrote, the technology of the jet engine that would evolve some 140 years later.

Countless other efforts followed. Each advanced by inches the development of internal combustion of vaporized gas in a repetitive, reliable manner. Then, in Paris in 1860, Etienne Lenoir received a patent for his gasoline-burning engine with two opposed cylinders, similar to the duplex, or compound, engines of steamers. As one piston began its downstroke, it sucked in a gas/air mix that was immediately sparked, forcing the piston the rest of the way down. The flywheel's momentum brought the crank around. This brought the opposite piston down, forcing the other one up to blow out the exhaust. It was smooth and quiet, and it was as powerful and three times as fuel efficient as contemporary steam engines of equal output. By 1864, there were 130 of them running in Paris alone.

▲The Wallis Tractor Company was located in Cleveland, Ohio at the time it produced the Bear. Founder H.M. Wallis was a relative of Jerome Increase Case and Wallis served for a time as president of J.I. Case Plow Works in Racine, Wisconsin, founded in 1876 to build plows (this was not the same company and had no relation to the J.I. Case Threshing Machine Company, also in Racine, except to guarantee confusion over use of Case's name). In 1919, Wallis Tractor Company merged with J.I. Case Plow Works. The Plow Works was bought by Massey-Harris in 1928 and the J.I. Case name alone was sold to J.I. Case Threshing Machine Company almost immediately. This cleared up the confusion.

◄The Wallis Tractor Company of Racine, Wisconsin, first showed the Bear in its 1902 catalogs. By 1914, Wallis was owned by J.I. Case Plow Works but in February, 1927, both were purchased by Massey-Harris Company, Ltd., of Toronto, Ontario, Canada. Tractor manufacture then continued at the J.I. Case Plow Works factory in Racine.

In Cologne, Germany, in early 1862, Nicolaus Otto was a distracted thirty-year-old. Working as a salesman by day, Otto apprenticed himself to a machinist, and he worked nights on his own projects in his master's shop. He had his own ideas that were based on Lenoir's machine. The noise of his ideas kept the neighbors awake.

Experimenting first with alcohol, then with a variety of petroleum spirits, he fabricated a kind of vaporizing carburetor that heated alcohol and pumped it into a cylinder. He labored and struggled for years. His intention was to get away from a fuel

▶ *The rear drive wheels were 30x84in. Despite being produced so early in the gasoline engine age, this was an extremely advanced tractor. It offered power steering, individual turning brakes, a spring-loaded clutch, an enclosed three-speed transmission, and an all-speed governor. Many of these features would not appear on other makes for decades.*

▲ *Daniel Best produced crawler tractors he called Tracklayers. Steering was accomplished by a differential incorporating steering clutches that slowed the track inside the turn while speeding the track outside. As the dirt marks indicate, the tractor could turn nearly inside its own length using this type of steering.*

source that was fixed to a building. He knew that taking his fuel from the gas lines that provided heat and light for the building limited the usefulness of his ideas. And the noise of his labors continued to keep the neighbors awake. As he wrote in his first patent application, his ultimate goal was to produce an engine "to propel vehicles serviceably and easily along country roads, as well as prove useful for the purposes of small industry."

Patent denied.

The Royal Ministry of Commerce of the Kingdom of Prussia had already patented vaporizing carburetors; however, nothing in the ministry's rejection spoke of self-propelled, self-contained vehicles. Otto convinced his master, Michael Zons, to construct a Lenoir-type engine. Then Otto tinkered and puttered. He altered and modified. Still, he kept the neighbors awake, and almost by mistake, he discovered something significant.

Experimenting with his fuel-and-air mixture, he allowed

▲The Best 25 tractor was produced from 1918 through 1920. The Model B engine was available without the tractor at a cost of $832 and was used in saw mills and to run harvesters. This Model B engine was Best's own design and produced 25hp at 800rpm, 12hp on the drawbar, and was capable of pulling 2500lb in low gear. This is a 1918 model.

▲*Best's Model 25 was the test model for what would become the Model 30 and 60. The 25s introduced an enclosed transmission and steering clutches. Previously, all this machinery had been exposed to outside dirt. In addition, the Model 25 was fitted with a gear-drive oil pump. The oil was pumped into a trough in the sump pan. As the crankshaft rotates, the connecting rod dipped into the trough and oil was splashed to where it was required.*

more of the mix to be sucked into his test cylinder than was usual. He wondered what would happen if he turned the flywheel back one more turn, to drop the piston as far down its travel as it would go. He reasoned that this would allow even more mixture into the cylinder. When he released the flywheel, the mixture ignited violently, and the flywheel rotated several more times on its own rather than just the usual one revolution he had achieved in all his previous attempts.

Back up the flywheel then. Let the fuel in. Release the flywheel. Spark, explosion, and several revolutions. And again. Un-

◄*Best used an Ensign-designed 1.5in carburetor and a Bosch or Splitdorf magneto on the Model B. The Model B engine was an in-line four-cylinder 4.625x5.25in engine with a total capacity of 302ci. This rare tractor is owned by Jerry Clark of Ceres, California.*

til he knew he could repeat it every time. Until he began to imagine it running longer than just those several revolutions. What about running continuously?

So again he repeated the process, and then he concluded that he needed a new timing gear. What he had found was that the Lenoir design did not allow for the extra stroke that would let in the additional fuel that seemed to turn the engine crank more times.

Otto allowed a full downstroke of the piston to fill the cylinder and a full upstroke for compression, then let it spark! A full downstroke of combustion power was the result. Then it began again. His flywheel brought the piston back to the top, to push out the burned fuel, and then back down to suck in a fresh mixture. Confidently, he and Zons created a new engine. But this time, showing a greater faith in his own conclusions, he built

▲*This 1920 Fordson was coupled to a 1928 Belle City Corn Picker-Husker. Belle City Manufacturing was based in Racine, Wisconsin, and this example was sold through Port Huron Machine and Supply of Lincoln, Nebraska. With the spout, the entire machine stood 97in tall, 190in wide, and 160in in overall length from front to rear.*

four cylinders. Just after the new year began in 1863, Nicolaus August Otto's first four-cycle, four-cylinder, internal-combustion engine woke the neighbors. It was not perfect. It ran unsteadily. It quickly wore out seals and bearings. The engine fired strongly at first, but the cycles afterward were rough and uneven. It was months later that he recalled James Watt's application of von Guericke's atmospheric principles. The piston of Watt's steamer was pushed back by the reintroduction of atmospheric pressure to the vacuum created by the escaping steam. For Otto this meant opening the intake valve earlier.

Another patent application. Another denial in Prussia. The atmospheric principle was nothing new in Germany, and Otto's use for it was not new enough. But in London, in Paris, and in Brussels, differences were recognized and patents were granted.

▲ *Fordson tractors were manufactured by Henry Ford in Dearborn, Michigan. Ford was unable to use his own name on the tractor because entrepreneurs in Minneapolis employed a man named Ford just to use his name on their inferior products. Henry produced his Fordson, named after Ford & Son, from 1918 through 1928.*

Now, short on money but still long on confidence, Otto found an investor, Eugen Langen, and a factory site on which to expand his experiments. N. A. Otto & Company was opened in 1864.

Work continued, and Langen, a technology institute graduate and the son of a wealthy banker, actually contributed the key to the Otto engine's success. He proposed attaching a freewheeling gear that alternately was driven by and then drove the main crankshaft. It was patented in 1866, and it was shown in May 1867 at the Paris World's Fair. In a building filled with Lenoir gas engines powering all kinds of demonstrations, Otto's new tall, single-cylinder, Greek-column-like, chattering, four-cycle gas engine consumed only one-third as much gasoline as the Lenoir. In Etienne Lenoir's hometown, Otto and Langen received the exhibition's gold medal from Napoleon III.

Success followed the gold medal, and sales followed the success: forty-six engines sold in 1868, eighty-seven in 1869, 118 in 1870. In 1871, Otto and Langen accepted 515 orders, many of them from North America.

Over the next several years, Otto continued to muse and develop. He perfected his fuel/air mixture and spark so as to burn the full length of the cylinder because the piston pulled the flaming brew to the bottom. This consumed the last unburnt fuel of the previous cycle as well. The longer stroke led to the development of Otto's high-compression engine, introduced in 1876 at his new factory in Deutz.

It would be another ten years before Gottlieb Daimler and Karl Benz would incorporate Otto's engine into their first automobile. What they needed was a smaller power plant, one capable of higher speed—800, even 1000rpm, instead of Otto's few

▲*This 1920 Minneapolis 12-25 tractor strongly resembles the Waterloo Boy Model N 12-25hp tractors produced by Waterloo Gasoline Engine Co. at the same time. Minneapolis' version was to be the second-ever tractor test at the University of Nebraska, in March 1920. But it was snowed out. Retested in May, it recorded 16.3 drawbar and 26.2 belt pulley horsepower. It is owned by Walter and Bruce Keller, Kaukauna, Wisconsin.*

hundred. Otto had experimented with advancing the ignition timing as engine speed increased. Daimler licensed Otto's engine, and he financed its high-speed improvements himself.

Subjected to innumerable patent infringement suits from the mid-1870s on, Otto's engine entered the public domain in June 1890 when the Supreme Court of Justice in Germany canceled his last patent. The court knew it was too valuable to remain under control. Otto knew what that meant.

He wrote to his wife, Anna: "Now the dance will begin in earnest. Everybody and his brother will be making engines."

Within months of Otto's engine becoming available, dozens of makers throughout Europe and North America offered for sale small, single-cylinder portable, transportable, and stationary gasoline engines. These produced as little as 1hp and as much as 65hp out of single-cylinder engines and 120hp from twins. The smaller engines saw steady acceptance for use around the farms, powering everything from water pumps and corn shellers to washing machines, butter churns, and cream separators. As had been done with stationary and portable steam engines, permanent applications were set up. These were separate "power buildings" on the farm, with rooms devoted to the cleaner or the dirtier tasks, all functions powered by belts driven off of ceiling pulley shafts belted up to the engine.

At almost the same time, engineers and inventors began to adapt gasoline engines to steam traction engine frames and running gear. The development of gasoline traction engines flourished throughout North America more quickly even than Europe. In 1900, while fewer than seventy firms in North America still produced steam traction engines, more than 100 manufactured internal-combustion gasoline engines, many of these designed for use in tractors.

The Charter Gas Engine Company in Sterling, Illinois, went into business in 1882, striking up an agreement with a German immigrant engineer, Franz Burger, to produce coal oil, as

◄*Ford powered its compact, 2,700lb tractor with an in-line four-cylinder engine with 4.0x5.0in bore and stroke. Power output was rated at 9.3 drawbar and 18.2 belt pulley horsepower at 1,000rpm. The tractor and Belle City corn picker were restored and are owned by Kermit Wilke of Wilcox, Nebraska.*

▲*The Minneapolis Threshing Machine Company's Model 12-25 measured 68.5in to the top of the throttle, 76in wide, and 198in long overall (about 48in longer than the similar-looking Waterloo Boy) on its 99in wheelbase (about 8in longer than the Waterloo). Front wheels were 6.0x38 while the rears were 12x56in steel with 2in tall cleats. It used automobile-type steering.*

well as gasoline-fired engines, an improvement over natural gas or coal gas versions. Burger worked for the next decade producing designs that founder John Charter patented and produced. Burger's most important design, for the liquid fuel engine in September 1887, opened the way for truly portable—and transportable—power. Two years later, Charter introduced its first self-propelled gasoline traction engine made from adapting its 10 to 20hp, 160rpm engine to a Rumely steam traction engine frame and running gear.

The next year, 1890, George Taylor, a Vancouver, British Columbia, engineer, earned a patent for his worm-drive walking plow. It resembled a wheelbarrow. Its single-cylinder engine was placed above the plow moldboard. On the ground behind the plow, a screw, operated by a chain driven off the engine crankshaft, propelled the machine.

A year after that, in 1891, William Deering & Co. introduced a 6hp tricycle mower that met with some success. In 1892, John Froelich of Waterloo, Iowa, paid $1,050 for a Van Duzen No. 12 vertical single-cylinder 20hp engine that Van Duzen manufactured in Cincinnati, Ohio. Early gas engines often used open flame ignitions, but by this time, Van Duzen Gas & Gasoline Engine Co. had progressed to the hot tube-type ignition. Af-

▲*The Minneapolis used a four-cylinder vertically—and transversely—mounted engine, different from Waterloo Boy's horizontal two-cylinder. Bore and stroke measured 4.5x7.0in. The 6,600lb tractor was produced from 1920 through 1926. This example is owned by Walter and Bruce Keller of Kaukauna, Wisconsin.*

ter much effort, Froelich fit the Van Duzen to a heavily modified Robinson & Co. "Conqueror" steam traction engine chassis, and he set to work.

Froelich worked in fields for seven weeks. He hauled his Case threshing machine behind his engine from one custom threshing job to another. He founded the Waterloo Gasoline

▲ This was the Stockton-built version of Holt's 5-Ton model, a name which designated its gross weight, in this case, about 9,700lb. Only 212 of these machines were built in Stockton. This crawler was also referred to as the 25-45hp Caterpillar. This Caterpillar was restored by and is part of the collection of the late Joseph Heidrick, Sr., of Woodland, California.

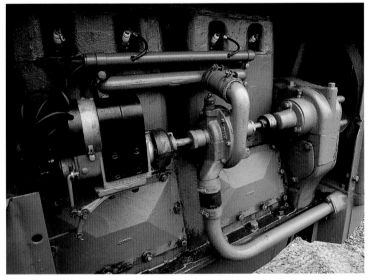

▲ Holt's in-line four-cylinder engine has a bore of 4.75in and a stroke of 6.0in. At 1,050rpm, the engine produced just about 33hp from the drawbar. Holt fitted a Kingston carburetor and an Eisemann Model 64 magneto. The engine ran on gasoline. This M-29 was built in 1923.

Traction Engine Co. in 1893. This was an effort to capitalize on his widely publicized accomplishment; now he would build tractors for sale. But he only completed three similar machines in two years, and his partners reorganized the business.

By the end of 1892, J. I. Case Threshing Machine Co. had also tried an experimental gas tractor. Case was the first of the major steam traction engine and threshing machine manufacturers to enter the market. Designed by William Patterson, the prototype even bore his name as well as the J. I. Case logo on the canopy over the engine. Fitted on the chassis of one of Case's smaller steamers, the Patterson was a horizontal, two-cylinder, four-cycle engine.

Before the end of the century, several others made first efforts, joining the market as existing steam traction engine makers. These included the McCormick Co., Huber Manufacturing, and Kinnaird-Haines (makers of the well-known Flour City threshers).

Within the first decade of the twentieth century, the market was flooded with old names as well as new makers trying new products. Charles Hart and Charles Parr had begun experimenting with gasoline engines as college students in 1895. In 1902, they founded a manufacturing company in Charles City, Iowa, and began producing two-cylinder 15 to 30hp tractors in 1902. In 1903, fifteen were sold and Hart-Parr was established as the first manufacturing company in the United States that produced only tractors. In fact, their sales manager, W. H. Williams, claimed the first advertising use of the word "tractor" in 1906, using it to shorten the expression "gasoline traction engine." In fact, the word had appeared in a U.S. patent application, #425,600, in 1890, issued to George H. Edwards of Chicago.

In 1902, in Stockton, California, John Kroyer designed and developed his Sieve Grip tractor at his Samson Iron Works. The Samson Sieve Grip used a unique style of wide, open-tread steel wheels that combined good traction with minimal soil compaction. The tractor was built low to the ground and was introduced with a single-cylinder engine and upgraded with a smooth-running four-cylinder.

At this same time, H. M. Wallis, a son-in-law of Jerome Increase Case, was president of his own firm, the Wallis Tractor Company, located in Cleveland, Ohio. In 1902, Wallis introduced its Bear model, a ten-ton, giant tricycle tractor that remained in production for another decade. The Bear's in-line, upright, four-cylinder engine displaced 1,481ci and drove not only the two huge rear driving wheels but also a power steering system. Maneuverability was further aided with independent rear-wheel brakes. In ten years of production, reportedly, only nine Bears were manufactured despite the model's many advances over both steam engines and other gasoline tractors.

In 1912, Clarence M. Eason joined the engineering staff at Wallis, working under Robert O. Hendrickson, the chief engineer. Together, they developed the unit frame system out of boilerplate. This was the first example of using the engine to serve as part of the structure of the tractor. In 1915, their Wallis Cub appeared with its transmission enclosed in the same U-frame assembly as the engine. In 1919, with the Model K, Hendrickson extended the U-frame to the full tractor length, including the drivetrain bull gears. By that time, however, Eason had left Wallis to work for Hyatt Roller Bearing, a company that would soon supply Henry Ford with bearings and ideas about unit frame tractor construction.

In Titusville, Pennsylvania, chemical engineer A. Fasenmeyer discovered a method to remove gasoline from the natural gas that was a by-product from oil wells. Fasenmeyer compressed this gas and then slowly cooled it, passing it through pipes placed in water tanks, similar to the condenser systems used by Savery, Newcomen, and Watt. Fasenmeyer obtained the product known as "casinghead gasoline." This process, after further improvements, became known as fractioning, gasoline being one of the fractions.

Three thousand miles west, also in 1904, Benjamin Holt began experiments with gasoline engines in crawler tractors. His work producing and selling steam-powered crawlers convinced him of the continuing market for crawler tractors for the soft sandy soil of central California and for use around Sacramento, where several rivers converged to produce rich, moist, spongy loam. By 1908, his first Model 40 25hp gasoline-engine "Caterpillars" were offered for sale out of his works in Stockton, California.

At the beginning of 1906, International Harvester Co. (IHC), completed its first tractor, a single-cylinder 20hp machine using a hit-or-miss ignition and friction drive. The friction drive was replaced almost immediately with a two-speed forward/one reverse sliding gear transmission that used a friction clutch and a final drive gear. At about the same time, IHC's chief engineer Edward Johnson completed development work on twenty-five prototype 8hp, two-cylinder, air-cooled, chain-driven automobiles. These were meant to be sold to farmers through IHC's tractor and implement full-line stores. Despite IHC's early successes with trucks, the IHC car never materialized. International tractors, however, were quite successful. From then on, the company devoted its time and efforts to the areas of its expertise.

▲ Holt used a three-speed transmission with a top speed of 5.5mph. Holt replaced this model with the 5-Ton which remained in production until Holt Manufacturing Co. in Peoria, Illinois consolidated with Best Manufacturing of San Leandro, CA and together formed the company known as Caterpillar Company in September 1925.

In 1906, car-maker Henry Ford entered the tractor market with a 24hp, four-cylinder machine based on his Model K automobile. While it was not a perfect machine, it convinced him that if anyone could build a perfect tractor, it would be Henry Ford. Out of that conviction developed a four-decade-long competitive battle that humbled and destroyed countless manufacturers and substantially strengthened and improved every surviving tractor maker and its products.

Beginning in 1908, in an attempt to bring more information to the farmers, the Winnipeg Industrial Exhibition began to include a "Light Agricultural Motor Competition." This served not only to show farmers the capabilities of various makers, it also put the makers at risk. If their tractors were unreliable or not strong enough, they would break before an audience. Eight tractors were entered in 1908. One withdrew before the competition began, and a four-cylinder 20hp Universal broke during testing. But three tractors by International Harvester and a single entry each from Transit Thresher Co., Marshall Sons & Co., Ltd., and

▲ *Both Ben Holt and Dan Best first built crawler tractors with tiller wheels mounted far out front for balance and steering. Holt's individual track clutches and brakes accomplished most of the steering; engineers in the Stockton shops relocated an engine closer to the rear and found it balanced well without the tiller. Over the years, each manufacturer produced models comparable to the other and both devoted considerable resources to legal battles over patent infringement. This 1917 Holt 75 belongs to Don Hunter, Pomona, of California*

Kinnaird-Haines Co. went through the workout. The Kinnaird-Haines four-cylinder 30hp machine won on points. The competitions continued through the July 1912 exhibition. Entered into the competition were a Sawyer-Massey four-cylinder 22 to 45hp tractor and Canadian licensees from Heer Engine Company, with its four-wheel-drive tractors, and Canadian Holt Tractor Co., with its Model 40 Caterpillar, along with twenty-five other machines.

According to the U.S. Department of Agriculture, in 1910 53.7 percent of the U.S. population, some 49,348,883 individuals, lived in rural areas. Of those, a third were engaged in agricultural occupations. In Canada, 39.9 percent of the working population, 716,937 people, worked in agriculture. Of the more than 1.9 billion acres of total area in the United States, 47,097,000 were producing wheat alone, averaging 14.7 bushels per acre. In Canada, with 2.4 billion acres of total land, barely 9,945,000 acres were producing wheat, averaging a yield of 19.7 bushels per acre.

Most gasoline tractors sold in 1910 were four-stroke engines with hit-and-miss ignition governors to control their speed. A low voltage DC magneto, or in some cases a generator, powered the ignition while the engine was running. Models equipped with electric start used dry cell batteries.

The initial ignition efforts by gas engine inventors had used an open flame that was exposed to the combustion chamber by a slide valve. This moved at the proper instant to ignite the fuel mixture on top of the piston. It worked fine for engines with little or no fuel compression, but as engineers learned that more power was produced if the fuel mixture was compressed and then ignited, a hot tube system developed. In this technique, a tube containing platinum, nickel, or even porcelain was kept hot by a flame burning inside it. The tip of the tube projected slightly into the combustion chamber. But both of these systems worked only for constant speed engines, in mostly stationary applications. When engines began to move, or when various loads required different speeds, low-voltage magneto-type systems and

high-tension coil, distributor, and spark plug systems evolved.

In the high-tension system, electric energy was increased so high that the current jumped between the electrodes of the spark plug, igniting the air and fuel mixture in the cylinder. In the lower-tension system, the spark was made inside the cylinder by bringing two electrodes together. Upon separation, the two electrodes sparked, igniting the mix. The current was produced either by chemical action that occurred in a dry cell battery or by electromagnetic action created by a dynamo, or magneto, and an induction coil.

In 1910, some 4,000 tractors, both gasoline and steam, were sold throughout North America. By 1920, 203,207 gasoline (or kerosene) tractors were manufactured, of which 29,163 were destined for export. By the end of 1910, the manufacturers had de-

▲ *The first fifty of the Model D tractors were fitted with a flimsy "ladder" style front axle. This was assembled from a variety of pieces welded together for what was presumed to be greater strength. After the fiftieth Model D, Deere & Co. went to a stronger solid two-piece casting that was welded together.*

◄ *Deere's horizontal two-cylinder engine measured 6.5x7.0in bore and stroke. In University of Nebraska tests, the Model D produced 22.5 maximum drawbar horsepower at 800rpm, with a peak of 30.4hp available on the belt. Earliest Model D's used the 26in diameter spoked flywheel.*

veloped engines that would run on cheaper kerosene, as well as on gasoline. The Canadian Heer Engine Co.'s 25hp opposed two-cylinder produced 17.5hp in the Winnipeg tests. A Rumely Oil-Pull produced 29.5hp in the same tests. Both ran on kerosene.

The sales competition between steam traction engine and gas tractor makers was fierce, but in organized events like the Canadian Industrial Exhibitions, the steam traction engine makers got clear glimpses of the future. In the 1912 competition, of twenty-eight tractors entered, only four were steamers and even the best of these, a J.I. Case 36hp model, was outperformed by a Rumely Type E two-cylinder and by an Aultman-Taylor 30-60.

In his 1917 book, *The Modern Gas Tractor: Its Construction, Operation, Application and Repair*, Victor Pagé's final chapter profiled the tractor designs for that year. In his opening paragraph, he summarized what the farmer could expect. "The most prominent feature noted in late tractor designs is the endeavor of builders to have light and strong tractors better adapted to general work on small- and medium-sized farms than the earlier heavy designs. The trend is unmistakably toward the small- and medium-weight machine, just as the trend in automobile designing is toward the medium-weight cars of moderate price. Tractor prices have been reduced and their use is increasing in all sections of the country."

The first gasoline tractors evolved from large, horizontal, single-cylinder, stationary engines. These were mounted on a supporting frame or chassis nearly identical to those used on steam tractors. By 1920, multiple-cylinder engines were used because they vibrated much less. The general arrangement of engine, chassis, and running gear was altered as a result. The steam tractors operated with their drive gears—and the engine parts—exposed to dust and the weather. Heavy, unfinished cast iron was used. By 1920, cut alloy steel gears ran in oil and operated somewhat like the gear-change systems on automobiles. Pressed steel in channels or I-beams had replaced the huge, heavy cast iron frames used on steamers and the early gas engine tractors. Cooling and lubrication, at first almost ignored, were largely—though not universally—accomplished by closed, pressurized systems.

It was in the preface to his book that Pagé hinted at the future: "It is not the writer's intention to underestimate the advan-

◄Tractor company equipment changes have evolved into the kinds of minutiae that confirm history for collectors. The four slots in each spoke of the steering wheel were only used in the 1923 model year. Three appeared in 1924, two were cut into the spokes in 1925 and in 1926, Deere jumped sequence and used solid bars. This is the 1923 version.

►This was the 32nd Model D tractor John Deere manufactured. Its engine was adapted from Waterloo Boy gasoline tractors which Deere & Co. purchased in 1918. However, while some of the Waterloo Boys ran on gasoline, Deere's Model Ds ran on kerosene. This rare example is owned by Lester Layher of Wood River, Nebraska.

tages and utility of the steam tractor; it has and still is performing work of great value. The gas tractor, however, in its modern forms, is able to accomplish everything the steam propelled type can do, and has important advantages the other construction does not possess. It does not require the services of a skilled engineer to operate, it has a wider range of action, is more independent of fuel and water supply in that it does not consume much liquid in cooling, and is more economical of fuel because it utilizes a larger proportion of the potential energy or heat units of the combustible by burning it directly in the cylinders."

As Pagé pointed out, many of the large gasoline tractors were smaller than some of the smaller steamers. That weight differential translated to benefits in two ways: First, it simply made the tractors more useful to smaller acreage farmers, and second, it made them less expensive to purchase and to own than the big steamers. With the development of even cheaper fuels—kerosene, diesel fuel, and the liquefied petroleum gases already on the horizon—the handwriting was beginning to show on the walls around the steam traction engine manufacturers.

▲The Fond du Lac kit required that the owner replace the rear wheels and tires with nine-tooth pinions. These engaged a ring gear mounted on the inside of the larger replacement steel wheels. A subframe was clamped on to the Ford frame and this extended the wheelbase nearly 2ft. The manufacturer claimed the conversion could be accomplished in 15-20min.

◄Don and Patty Dougherty of Colfax, California, restored and own this soft-top pickup tractor conversion manufactured by the Fond du Lac Tractor Co. of Fond du Lac, Wisconsin. Kits to convert the very popular Ford Model Ts were sold by a number of makers in the late teens and twenties for between $100 and $200. Thousands were sold but most were scrapped during iron and steel collection drives during World War II.

▲"The power and reliability of the Ford engine," Victor Pagé, a highly regarded automobile authority, wrote in his 1918 edition of The Model T Ford Car and Ford Farm Tractor, "makes it possible to use this light chassis for much heavier work than one would imagine it capable of. A variety of tractor attachments are provided by which the Ford chassis may be used for light agricultural work that would ordinarily be done by several horses, such as plowing, harrowing, cultivating, etc." But, he added, "A machine which is built up of a standard touring car and a tractor attachment cannot be expected to do the same kind of work that a specially designed tractor can do." He went on in later pages to introduce the Fordson tractor.

▲ *The Ford Model T engine was designed as a unit-type power plant. The engine, flywheel, and transmission are all within the same two-piece housing. This allowed Ford to ensure perfect alignment between the crankcase and the transmission shaft. Bore and stroke of the four-cylinder engine was 3.75x4.0. Power output was quoted at about 20hp.*

▶ *In the thirties, Model Ts were so cheap—at $10 to $50 each—that routine maintenance was performed only on the conversion kits; the Ts were run until they broke and then were replaced. The Fond du Lac was in competition with such companies as Convertible Tractor Corp., Curtis Form-A-Tractor Co., Handy Hank Tractor Attachment Co., Smith Form-A-Truck Co., and Uni-Ford Tractor Co., among others. More than fifty companies existed in 1920. This is a 1926 Model.*

▸*This 1929 10-20 TracTracTor was produced from 1929 through 1931. A total of 1,505 crawled out of the Chicago factory. No carrier idler wheel was used (a carrier idler wheel is a roller that holds up the track along the top of its travel between the sprocketed drive wheel at rear and the front idler). Three track rollers along the bottom kept the track in contact with the ground over any contour.*

▲McCormick-Deering offered the Model 10-20 in both wheeled tractor versions as well as crawlers. These were called the TracTracTors. This line was introduced in 1929 with this model. A prominent engineering and visual feature was its pair of track clutches under twin cones just behind the engine panel.

▼The 10-20 TracTracTor used the same engine as the wheel-type Model 10-20. A vertical, in-line four-cylinder 4.25x5.0in engine produced 15.5 drawbar and 24.8 belt pulley horsepower at 1,025rpm. By this time, International produced its own magnetos while still buying carburetors from Ensign. This crawler is owned by Mrs. Edith Heidrick, widow of the late Joseph Heidrick, Sr., of Woodland, California.

◀Bore and stroke of the in-line four cylinder engine was 3.375x4.0in. In its tests at the University of Nebraska, the Cat Model 15 produced a maximum 16.2 drawbar horsepower and 20.4hp brake horsepower at 1,500rpm. Ignition was by an Eisemann Model GV4 magneto and Caterpillar fitted Ensign carburetors on the Fifteens.

▼The tall lever is the master clutch while the two matching shorter ones operate left and right track brakes for steering. Ahead is the gear shift. Caterpillar used a three-speed transmission on the Model 15, with a top speed of 3.5mph. The 4,750lb tractor was capable of pulling a maximum of 3,105lb in first gear in its tests at University of Nebraska.

▶▶Introduced in 1929 and offered through 1932, the Model 15 sold for $1,450 at the factory in Peoria, Illinois. It used Caterpillar's own four-cylinder engine. This 1929 example was restored and is owned by Frank Beitencourt of Vernalis, California. Behind the crawler is Bettencourt's John Deere-Killifer Model 7MKO-02 disk.

◂The 1935 Model B measured 76in tall, 79in wide, and 119in in length on an 81in wheelbase between its front and rear steel wheels. Deere used a Fairbanks-Morse DRV-2B magneto and a Schebler DLTX10 carburetor. The Model B weighed 3,275lb.

▴Deere's Model B engine continued on the Waterloo heritage of two-cylinder horizontal engines fitted with valves-in-head. Bore and stroke were 4.25x5.25in. Drawbar horsepower tested out at 11.8 while belt pulley output was 15.1hp at 1,150rpm. Four speeds allowed for a maximum of 6.4mph yet it was capable of pulling 1,728lb in first.

▸▸This was the very first Model B that was produced, number B1000. Introduced in late 1934, it was designed as a smaller, companion model to the big Model A (introduced in April 1934). It was intended to fill the shoes of two horses on any farm. This one was restored and is now owned by Walter and Bruce Keller of Kaukauna, Wisconsin.

Chapter Three

The Nebraska Tractor Tests

A Bogus Ford Brings Accountability To the Tractor

W. BAER EWING HAD AN IDEA FOR A TRACTOR, but he needed help. He didn't need a designer or an engineer. He needed a name, and he knew the name he wanted. When he found him, Ewing hired a young man named Paul B. Ford.

Ewing was a Minneapolis entrepreneur. Involved with the Federal Securities Company in Minneapolis, he had sold bonds and securities in the Power Distribution Company. Electric power was being introduced throughout major cities, offered to businesses and home owners alike. Rural electrification was decades away, but it was anxiously awaited in the urban areas. Investors were needed to fund development and installation, but when those bonds came due, Ewing came up short and many of his investors lost out.

Somehow, Ewing emerged unscathed. He recognized a new need in the marketplace and a new opportunity for investors. Presumably he was aware from newspaper accounts that automaker Henry Ford was interested in the farm tractor. Ewing sought to capitalize on Ford's name and the success of his automobiles. With his new assistant, Paul Ford (no relation to Henry Ford), Ewing founded the Ford Tractor Co., incorporated in South Dakota probably sometime during 1915. According to C. H. Wendel in his *Encyclopedia of the American Farm Tractor* and other sources, Paul Ford knew nothing of farm tractors, but that didn't matter to Ewing who began promoting him as the designer of Ewing's new tractor, a machine actually designed by a character named Kinkaid.

To finance the tractor's production, Ewing sold stock through his own company, Federal Securities. Then he sold tractors to unsuspecting farmers. The company went broke in 1916, but Ewing managed to hold onto both the patents and designs, and

▲*The Fordson pulls a Ferguson-Sherman two-bottom plow attached to its Ferguson Duplex hitch. When car and tractor-maker Henry Ford would not give Irish engineer Harry Ferguson the respect he felt he deserved, Ferguson went into business with two brothers in Detroit to produce his plows and hitches. These were meant for use on—and often sold with—Fordson tractors.*

▶*Palmer Fossum's 1926 Fordson sits ready for the next morning's work. A homemade tin-and-wood cab was simply bolted onto the fenders. The racket it made in vibration was louder than the Ford four-cylinder 4.0x5.0in engine. At 1,000rpm, the Fordson produced 12.3 drawbar horsepower, adequate to pull a two-bottom plow across Fossum's Northfield, Minnesota, field.*

to Paul Ford and his name. Ewing relocated the whole operation to southeast Minneapolis and opened his doors a second time to an unsuspecting marketplace.

His Ford Model B tractor (there was no Model A) sold for $350 and was rated at eight drawbar and sixteen belt pulley hp. He advertised that it was capable of pulling two 14in plows. It used an opposed two-cylinder Type M engine manufactured by the Gile Boat & Engine Co., of Ludington, Michigan. The few tractors produced were probably assembled just to show potential investors an actual product, to convince them of the company's viability.

Those who bought stock to capitalize the company never saw any return; those who bought one of the few machines that Ewing sold learned even more quickly of their mistake. In mid-1917, Ford Tractor Company was brought before a U.S. grand jury in New York as part of an investigation of Ewing's partner's involvement in other stock sales schemes. The partner was convicted and sentenced for conspiracy to defraud investors and, according to Wendel, Ewing relocated to Canada to organize yet another tractor company.

This kind of skullduggery was not uncommon in the first twenty years of gasoline tractor manufacturing. There were many other tractor makers organized solely for the purpose of making money, not tractors. One prototype would be produced, funds would be raised, and overnight the company was out of business, offices vacated, doors locked. The lone prototype had sold for cash to some unsuspecting victim.

One such victim, however, was in a position to do something about his misfortune. It didn't get him his money back, but almost single-handedly he cleaned up the tractor manufacturing business.

Wilmot Crozier, a progressive farmer

◂◂*The Gile-engined Ford 1919 Model B sold for $350 fully equipped out of its factory at 2619-2627 University Avenue S.E., in Minneapolis. Described as a two-plow tractor, it used bull gears to provide power to each of the two large front drive wheels. It incorporated automobile-type steering. Ironically, Gile Tractor and Engine Company produced its own rather innovative tractors from 1913 until late 1918.*

▸*The engine Baer Ewing selected for his Ford tractor was manufactured by Gile Boat and Engine Company of Ludington, Michigan. It was the Type M, rated to operate between 700 and 900rpm, and quoted as producing eight drawbar and sixteen belt horsepower. There's no way to prove it because, of course, its very unreliability brought about the Nebraska tractor tests.*

from eastern Nebraska, bought one of Baer Ewing's Ford Model B tractors in 1916. The machine broke down so frequently that he demanded a replacement and got from Ewing a 1917 model. It was no better so he simply gave up and parked it. He replaced it with a 1917 Bull tractor, and it too failed him. Soon after, in 1918, still desiring tractor power for his farm, Crozier bought a second-hand Rumely Oil-Pull that worked better than his expectations. Rated at three-plows, the Rumely routinely handled five with no difficulty through the soil on his farm.

Crozier, college educated and world traveled—he had served four years as a superintendent of schools in the Philippines—was angered by his experience and frustrated by his me-

▾*The Model B measured 60in high, 78in wide, and 124in long overall. Its front wheels—the driving wheels—were 54in in diameter. Steering was by the rear wheel only; there were no separate clutches or brakes for either front wheel. The drawbar could swivel as much as fifteen degrees to either side of center.*

chanical innocence. He had read a farm journal editorial that recounted experiences similar to his own. The article called on the farm tractor manufacturing industry to clean its own house or face government regulation.

Newspapers in rural areas and farm journals at this time were strong advocates for farmers' rights and needs. Strong editorial support led manufacturers to produce smaller, more maneuverable tractors early on; these same printed voices encouraged the creation and development of general purpose tractors capable of more functions than merely initial field plowing, preparation, and planting. The role of these publications in the development of farm tractors was significant.

According to R. B. Gray in his 1975 book, *The Agricultural Tractor: 1855-1950*, as early as 1915 there was a growing interest in a federal government-established national testing station for farm equipment. Whether the Bureau of Standards would operate it or the Department of Agriculture was one of many ques-

▲The front wheels were 54in in diameter. The entire tractor measured 60in tall, 78in wide, and 124in overall length. A Bennett Model EB carburetor and International Harvester's Model E4A magneto were standard equipment. This tractor is being restored by University of Nebraska agricultural engineering students Jeff Hays and Brent Smith.

tions to be answered. The American Society of Agricultural Engineers advocated formation of a Bureau of Agricultural Engineering within the Department of Agriculture. All that was needed was funding.

Crozier, elected to a two-year term to the Nebraska State Legislature in 1919, approached several of his fellow legislators with the idea of a testing system, carried out by impartial judges. This would evaluate any tractor that was offered for sale in the state of Nebraska. The results would be public knowledge; if the tractor failed—or passed—everyone could know it. Surely a manufacturer that knew its tractors would not succeed the tests would not offer them, and therefore, no farmer in Nebraska would have to go through Crozier's experience again.

The Department of Agricultural Engineering at the University of Nebraska was chosen as the suitable testing facility, and on July 5, 1919, it was law. No new tractor could be sold legally in Nebraska without a permit, issued by the State Railway Commission. The permit, certified by a board of engineers, would

▲ *As significant as Henry Ford's Fordson was in farm tractor history, this Ford tractor built by a Minneapolis manufacturer to take advantage of the Ford name started a revolution in tractor manufacturing. W. Baer Ewing named the machines after an employee, Paul B. Ford, and produced them in Minneapolis. Nebraska tractor tests began due to the poor performance of these tractors. Ironically, this one is owned by the University of Nebraska Tractor Testing Museum. The museum is operated by its founder and former test director, Lester Larsen.*

prove the tractor had substantiated the claims of its manufacturer by a series of standard tests and procedures.

L. W. Chase, head of the department at the Lincoln campus, had also been actively involved in the Winnipeg Agricultural Motor Contests. He hired Claude K. Shedd from Ames, Iowa, to be engineer-in-charge of testing. Through the winter, the necessary procedures were devised and set up. Little money was available, and a small building was quickly erected, and equipment put in place. On March 31, 1920, the tests began.

Ten days later, a Waterloo Boy Model N 12-25 passed the first Nebraska tractor test.

Tractor makers with nothing to fear applauded the tests nearly as loudly as did Nebraska farmers. Tractor makers like Baer Ewing wisely stayed out of Nebraska. The eighteenth test performed at the university facility was Henry Ford's own tractor, called the Fordson because, through Ewing's shrewdness, Henry Ford did not own the rights to his own name for farm tractor manufacture. He called his Detroit company Ford & Sons.

▶▶ *Early Waterloo Boy tractors used the slack-chain steering mechanism. This allowed the front axle to pivot back and forth against the coil springs at the ends to cushion shock to the operator's hands on the steering wheel. It also required some advance thinking to take up the slack before a tight turn.*

◂Tested at Nebraska University's main campus in Lincoln, the Model N used Waterloo Gasoline Engine Company's horizontal two-cylinder engine with 6.5x7.0in bore and stroke. This engine style set a precedent with Deere & Co. that would continue from the time Deere bought Waterloo in March, 1918 until engines changed in 1960.

▴The Waterloo engine produced 15.9 drawbar horsepower and 25.9hp on the belt pulley. The Model N was introduced in 1917 and remained in production until 1924 when Deere and Company brought out its Model D. A two-speed transmission enabled the Waterloo to pull 2,900lb at 750rpm in first gear. This 1920 example is owned by Walter and Bruce Keller of Kaukauna, Wisconsin.

▸▸Once Nebraska Legislator/educator Wilmot Crozier succeeded in getting tractor tests mandated by state law, the legitimate companies lined up with their products, anxious for approval. The first tractor tested was a Waterloo Boy Model N similar to the one shown here. It passed the tests.

▲ Allis-Chalmers specified the Kingston Model L carburetor and an Eisemann GS4 magneto. Tested at the University of Nebraska, the Allis four-cylinder 4.75x6.5in engine produced 33.2 drawbar and 44.3 belt pulley horsepower at 930rpm.

▸ Original and unrestored, Conrad Schoessler's 1929 Model E needs only its seat to be complete. Introduced in 1919 as the Model 18-30, it was improved and uprated to 20-35 by 1929. It remained in production through 1930. About 2,155 were sold.

▸▸ The Model E weighed 7,095lb for testing yet pulled 4,400lb in the lower of its two gears. While the tractor was introduced at a selling price of more than $2,000, the fierce competition generated by price wars between Henry Ford and International Harvester reduced the price of all machines. By late 1928, the Model E 20-35 was selling for $1,295.

Chapter Four

Development of the Diesel

The Power and the Glory

The engine's German inventor had it in mind to call it the Delta engine, or possibly, "The Excelsior." On paper, he referred to it as the "Rational Heat Engine on the Diesel Patent," but while he was agonizing over its name and fretting so many other problems in early October 1895, his wife, Martha, said to him, "Just call it a Diesel engine."

Rudolf Diesel, a German born in Paris in 1858, had an idea for the engine while he was still a teenager. His higher education included several years at the School of Industry in London when his family lived there. He had earned a degree in engineering from Munich's Polytechnikum before he was twenty. Exposed to the theories and engines of Thomas Newcomen, James Watt, and Nicolaus Otto, Diesel challenged himself to invent a heat engine that was more powerful and more efficient than anything achieved by those three men whom he admired most.

Out of school in 1880, Diesel's first job was in a plant owned by Carl Linde, his former professor of theoretical mechanical engineering in Munich. Linde had invented a machine to make artificial ice using pressurized ammonia.

Working for Linde furthered Diesel's practical education. Gaseous ammonia can be condensed into liquid at about 50deg Fahrenheit under six to seven atmospheres—nearly 100psi—of pressure. Boiling it back to a gas consumes a great deal of heat out of, for example, a water reservoir. Simply put, returning liquid ammonia to gas will chill water to ice, condensing water to its extreme. This, of course, is exactly the opposite effect of a heat engine.

Diesel imagined that an engine might be made to work by compressing ammonia under perhaps fifty atmospheres. This could be done by cooling it—compressing it—rapidly through the action of a piston in a

▲*The Farmall M was introduced in August 1939 and became one of International's most successful and largest-selling tractors by the time it was discontinued in 1952 when it was replaced by the Super M series. In all, 279,821 of the Model M were produced. The electric starter and lights were optional in 1940, as was the power take-off and belt pulley.*

▶*The Farmall diesel Model M was announced in 1940 with production beginning in 1941 and continuing through 1954. It sold for $1,550. It was available in standard and row-crop fronts and by 1947, the high-clearance V-series was offered. A total of 8,298 gasoline MV and diesel MDV tractors were sold.*

cylinder. When the ammonia was reheated, the increased pressure would force that piston back. But liquid ammonia is a corrosive, and when his pipes leaked, the air around his "engine" became lethal. After five years of experiments (Diesel was nothing if not tenacious and thorough), he gave up on ammonia. By 1890, he had revised his idea and he began to conceive of a high-pressure air engine, the air heated quite hot by extreme compression. Into a cylinder filled with highly compressed air, a fine mist of fuel would be injected when the air was at its most compressed stage. An explosion should result, and this explosion would force the piston back down the cylinder. His paper on this theory, published in 1893, earned him his first patent.

Diesel's engine operated on the basis of Nicolaus Otto's four-cycle gasoline engine. But Otto's second stroke, the compression of fuel and air mix just before spark, became Diesel's compression stroke with plain air alone. Diesel's engine achieved six to eight times more compression than Otto's engine was intended to accomplish.

This high pressure heated the air at the top of the stroke so much that all Diesel needed to do was to introduce a volatile fuel mixture; no spark was necessary. The air was hot enough to immediately ignite the mix. It confirmed Diesel's theories. The cylinder's contents would explode through the third cycle, pushing the piston down as the combustion expanded the mixture in the cylinder. Exhaust followed on the fourth stroke, just the same as Otto's engine.

It required years of experiments to make it work. If he succeeded, his engine would provide significant fuel economy over gasoline engines. Because of this possibility, he was able to convince Machienenfabrik Augsburg to give him space in which to work. Industrialist Friedrich Krupp provided the financial backing with which to

▲ *International's in-line vertical four-cylinder engine measured 3.875in bore and 5.25in stroke. It used a Bosch fuel-injection system. In its 1941 University of Nebraska tests, it produced 25.4 drawbar horsepower and 35.0 horsepower on the belt pulley. In its lowest gear it was capable of pulling 4,541lb; in its highest gear, it would run more than 16mph on the road.*

▸ *The high clearance V-series yielded 28in of crop headroom under the drawbar. Overall the tractor stood 113in tall, 83in wide, and 141in long overall. Front tires were 7.50-20 while rears were 12.4-38s. This 1947 MDV is owned by Bob Pollock of Denison, Iowa.*

experiment. Diesel started in July 1893. By mid-1895, his first experimental engine was ready. With a crude fuel pump injecting a petrol mixture into his cylinder, which generated eighteen atmospheres of pressure, about 250psi, his theoretical engine fired at last. One time. He had wanted thirty atmospheres, but achieving it was impossible with the technology available at the time. It was a good thing. A pressure gauge that he had fitted to the top of the cylinder exploded from his engine's internal pressure. Pieces flew like shrapnel around his lab.

But it did prove that Diesel was right. His theory worked. It simply needed more work. Much more work, that is, because his engine would not yet run continuously. However, by 1897, Diesel had developed a fuel pump that operated at a pressure that was even higher than that of the air compressed within his cylinder. His new pump would blast the fuel mix into the cylinder at just the critical instant. This fuel-feed system lasted virtually unaltered until sometime in 1923, when Robert Bosch perfected his own high-pressure injection fuel pump.

Diesel's follow-up experiments with a multi-cylinder engine were a disaster. At the same time, Adolphus Busch, the St. Louis, Missouri, brewery owner, came to Munich to see Diesel. After thoroughly examining the engine and getting the opinions of other theoretical scholars and practical engineers, Busch watched some experiments. Then he signed a contract with Diesel in October 1897. Busch would produce diesel engines in the United States. Diesel, full of himself, asked Busch for a license fee of $1 million! Busch never blinked. And in 1898,

▸▸ *There's plenty of work ahead for Frank Bettencourt's D-6. Behind the crawler is his freshly restored John Deere No. 8 four-bottom 16in deep-tillage plow. This plow configuration is known as a three-by-four, the fourth plow positioned off to the left of the three main plows. Bettencourt restored both the plow and crawler for his collection at Vernalis, California.*

▲The D-6 was a powerful choice for agriculture in California's central valleys. The six-cylinder engine of 4.5x5.5in bore and stroke operated at 1,450rpm and produced 49.4 drawbar and a maximum of 76.9 brake horsepower during its Nebraska tests in July 1949. The 18,805lb tractor pulled a maximum weight of 16,222lb, clearly demonstrating the efficiency of crawler tractors.

▼The starting engine transmits its power through a single-plate clutch to a sliding pinion which is engaged with the diesel engine flywheel gear by means of a hand lever. The pinion is automatically disengaged by centrifugal force acting on the pinion latches when the diesel engine begins to operate under its own power.

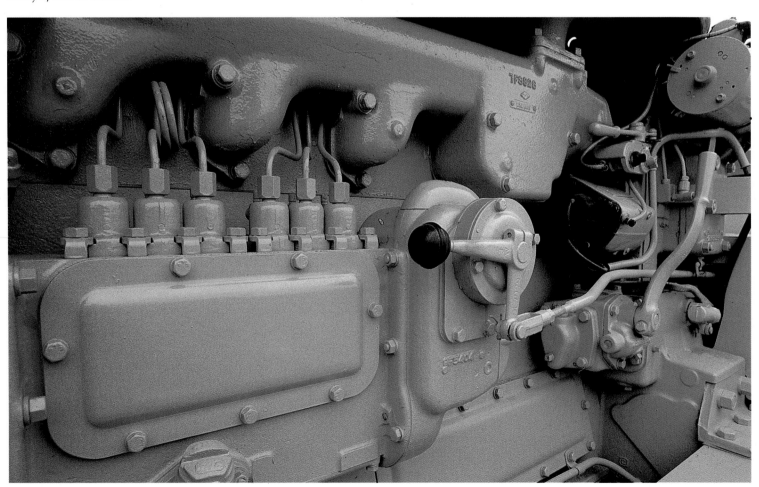

▲ *Caterpillar spent more than $1 million dollars during the late 1920s to develop and prove the diesel engine for use in mechanized agriculture. Introduced on the Caterpillar 60 chassis as the 1C-series in 1931, diesel power was not an immediate success. It was something new and as yet unproven. By the fifties, however, diesel was nearly all that Caterpillar sold. This is the 1952 Model D-6.*

Busch began manufacturing the first diesels in North America.

With his own newfound wealth to fund his work, Diesel began to develop practical uses for his engine. The French and the Russians had diesels in canal boats just after the turn of the century. Around 1904, Diesel began to imagine his engines adapted to locomotives and automobiles. By 1912, his large engines were in ships, and the Sulzer Brothers, a locomotive works in Winterthur, Switzerland, were first to experiment with a diesel to replace steam power. But those efforts failed several times. The mechanics came to call the Sulzer engine "The Foundry." With each failure, its heat melted the internal parts.

The world had come to understand what Diesel knew. His engine, when run alongside Nicolaus Otto's gasoline version, used as little as one-third the amount of fuel to produce comparable work. A group of American engineers came to Germany in early 1913 to invite Diesel to appear with his engine at the San Francisco Panama-Pacific Exhibition in 1915. He would sail to the United States on a diesel-powered ship.

But Rudolf Diesel had been psychologically troubled from the first by his failures. Then, following his indisputable success, he was plagued even more by the jealousy and treachery of competitors who tried to steal his invention or by critics who sought to revile it. And so on Monday, September 29, 1913, sailing on the North Sea from Hamburg to England to supervise the opening of his engine factory in Ipswich, Rudolf Diesel simply slipped over the side of the ship late in the night. His body was found and identification was confirmed days later. But because fishermen customarily did not take dead bodies from the sea, they let his body drift away.

It fell to Robert Bosch—maker of magneto sparking devices for airplanes, automobiles, trucks, and tractors—to substitute Diesel's high-air-pressure fuel-injection system with hydraulic injection of the mixture into the cylinders. Bosch's work commenced shortly after the November 1918 World War I armistice. A number of engineers and experimenters had begun working to develop a reliable fuel feed system for diesel engines. By late 1922, Bosch was seriously involved, and for the next four years, he and his engineers attempted to devise methods of blasting a microscopic drop of fuel as many as twenty times a second into a diesel combustion chamber against high back pressure.

By 1927, trucks from Benz, Daimler, and MAN (Machienenfabrik Augsburg had evolved into MAN around the time of Diesel's death) took to the road in a driving test that went from southern Germany through Czechoslovakia, Austria, France, and back into Germany. Bosch's pump on Diesel's engine was reliable enough that three more Benz diesel trucks be-

▲ *Standard equipment on the WD-45 included the automatic traction booster, two-clutch power control, a five-way hydraulic system, a four-speed transmission, power take-off, individual foot-operated rear wheel brakes, an adjustable hydraulic shock-absorber seat, fenders, twelve-volt electric starter, and lights.*

◄ *Allis-Chalmers introduced the WD tractor line in 1948 with a four-cylinder 226ci high compression gasoline engine available as a standard, adjustable wide front end, and both a dual-wheel and single-wheel row crop front ends. In 1955, Allis added its new overhead valve, six-cylinder 230ci diesel.*

gan making shuttle runs between various Bosch factories in Europe. Another one was shipped to New York in late 1927 for deliveries in the East. Bosch delivered its one thousandth diesel engine fuel-injection pump before 1928. At that point, it was ready to gear up for the mass production that it believed the world's industry would demand.

Diesel engines did make it to San Francisco for the Panama-Pacific Exposition. One, a McIntosh & Seymour four-cylinder produced 500hp at 164rpm and won grand prize at the exhibition. Another, displayed by the Danish licensee Burmeister & Wain, operated a generator in the Machinery Building, not far from where Best Tracklayer and Holt Caterpillar gasoline-engine tractors were shown. C. L. Best, son of founder Daniel Best, was running the Best company at the time, and he and his chief engineer, Oscar Starr, came frequently to the exhibition to see, study, and ask questions about the diesel engine. Its economy of operation convinced both men that it was necessary to get some version of it into a tractor.

Shortly after Adolphus Busch was licensed to produce diesel engines in St. Louis, another American visited Dr. Diesel during his early trips to Ipswich in 1911 and obtained production rights (Diesel had been granted several honorary degrees following the success of his engine). George A. Dow, a pump manufacturer

▲ *The new six-cylinder engine displaced 3.45x4.125in bore and stroke. Drawbar horsepower was measured at 30.5 at 1,625rpm, and a maximum of 43.3hp was measured on the belt pulley. In addition to drawbar and belt pulley power, an additional engine use was the "Minute-Quick" system that used engine power to adjust rear wheel track.*

►► *The standard front axle model weighed 4,785lb and measured 88.3in tall and 128in long. Front tires were 5.50x16; rears were 12x28, mounted on 12in steel rims. This 1955 Model WD-45 diesel is owned and used nearly every day by Clyde McCullough of Vail, Iowa.*

◀*The Marshall engine was a transverse-mounted, two-stroke, single-cylinder, horizontal diesel with a bore and stroke of 6.5x9.0in. Early models rotated counterclockwise; the final series tractors reversed crankshaft rotation. Marshall rated the tractor at 40hp at 750rpm. The radiator was split with the cooling fan placed in between the two halves. The Field Marshall measured 83in tall, 75in wide, and 113in in overall length with a 61in wheelbase. This compact, unusual tractor belongs to Walter and Bruce Keller of Kaukauna, Wisconsin.*

from Alameda, California, only a few short miles from the Panama-Pacific Exhibition site, received his license from the English makers in Ipswich.

Dow built twenty-eight diesels, all of them for marine installations, except the first one he built. That one Dow installed at his shops (and it continued running in daily use until 1952, forty years after its manufacture). These were three-cylinder 150hp engines. He manufactured diesel pumps until their price became too much of an issue with his customers, and then Dow quit rather than compromise his standards.

In the years following the 1915 exposition, Best and Starr maintained close contact with Dow. In late 1925, following the consolidation of Best and Holt operations under the name Caterpillar Tractor Company, Best, as the new Caterpillar chairman, found time to see Dow again.

Dow's diesel engineer was Art Rosen, a young mechanical engineer from the University of California. Rosen was available because Dow was quitting diesel manufacture. Ironically, Art Rosen had begun writing to the Holt company as early as July 1923. He hoped to interest Holt Manufacturing into applying the diesel to its tractors.

Meanwhile, farther up the east side of the San Francisco Bay, John Lorimer began building diesels for Atlas Imperial Company of Oakland in 1920. The Atlas diesels soon found applications in drag line and scoop shovels, fishing boats, and locomotives. The Atlas diesels were what sent Henry J. Kaiser in to see C. L. Best, Oscar Starr, and the young Art Rosen. "If you won't put Atlas Diesels in your Cat chassis," he reportedly said in frustration one day, "then I will!"

Kaiser had first pressured the tractor companies to adopt the diesel in the mid-twenties. John Lorimer and his son Ralph installed Atlas diesels in Caterpillar and Monarch crawlers.

Kaiser took three of the diesel-engine hybrids to a construction site on the Mississippi River. He discovered that the stationary-type Atlas engines were meant for use in an application that was far more structurally rigid and in an atmosphere much more environmentally controlled than what existed with an earth-moving crawler tractor. The heavy engines destroyed the Monarch and stressed a Caterpillar Sixty chassis to the breaking point.

But while Kaiser's experiences 2,000mi away from Caterpillar's headquarters in Peoria, Illinois, argued for a go-slow attitude, an event 12,000mi away injected fresh fuel against the back pressure of corporate resistance.

▲*There are a variety of ways to start the Field Marshall diesel. A special holder fit into the cylinder head to hold a kind of wick which was lit by a match. The decompression valve was opened, and the huge, heavy 25x5in flywheel could be levered around by a couple of strong men. Otherwise, another holder in the cylinder head would hold a blank shotgun shell.*

The operator would release the compression after the wick had burned into the engine. The shotgun shell would be fired with its holder. The tractor would almost inevitably start. It is not known how many times this could be done before the lower end of the engine needed rebuilding. This 1952 model belongs to Walter and Bruce Keller.

◂*Field Marshall tractors were manufactured at the Britannia Works of Marshall of Gainsborough in Linconshire, England, from 1949 through 1952. During those three years, more than 3,000 tractors were produced. The tractor weighed 6,500lb and was fitted with a three-speed transmission that provided a top speed of probably 7mph. Later models used a six-speed-plus-two-reverse gear box that gave a road speed of 11.3mph.*

Even in the late twenties, heavy plowing in many places in the world was still done with cable tackle systems. In areas where sugar cane, beets, and cotton were raised, two steam engines were hired, one placed on either side of the field.

But gas-engined Caterpillars had outperformed the steamers—even those with cable rigs—everywhere. J.& H. McLaren of Leeds or G. J. Fowler of St. Ives, England, the foremost promoters and practitioners of this technique, had generally conceded defeat to gas engine Caterpillar crawlers by this time. Now a cotton-plowing contest was proposed to pit a Caterpillar Sixty against a cable system. The Sixty would run lengthwise in the fields, and while it would leave head lines in the turns, it was expected to prove more efficient than the steam and cable plowing system.

So, in early 1927, Caterpillar shipped a gasoline Model Sixty and spare parts to Khartoum in the Anglo-Egyptian Sudan. Fowler did not show up with a steam tractor, but instead arrived with tractors that had Benz diesels in them. Caterpillar was beaten by a fifty-year-old system powered by a brand new engine. Peoria cabled its representative at the contest to go to Germany and buy one of the Benz engines. There were delays, but the engine finally arrived, and Caterpillar studied it.

Then Rosen and Starr designed the engine that came to be known as the D-9900, the Caterpillar diesel engine. They initially built three engines. When those prototypes were assembled and running, the whole board of directors came out to the lab. The board had learned that engineering had spent $45,000 to build the three engines. It couldn't imagine what could require so much expenditure, but it would take much more. Caterpillar spent more than $1 million on the diesel before ever marketing the first one.

In the Experimental Section of the engineering department in the late twenties and early thirties, the first diesels were installed in a Model Sixty chassis. From the beginning, methods of starting the diesel engine had to be made foolproof, like the engine itself, otherwise it would never sell. Gasoline "pony" engines were used almost from the start.

On some of the prototypes, engineers tried to start up the small gas engine with a flat belt. They ran it until it was hot and then piped the hot water through the larger diesel to warm it up. When it was hot, the valves were dropped, and the diesel was turned over for a period to heat it up evenly enough to stabilize the combustion process. Engineers learned that it was critical to quickly disconnect the belt between the gas engine and the diesel, because when that diesel started, it would tear the gas engine to pieces.

After Henry Kaiser's experience with the Atlas diesel flexing and overstressing the Caterpillar Sixty chassis, the engineering department knew it had to strengthen the frame and solve all the other engineering problems. A special large transmission was finally used for the diesel engine tractors. It was geared down more for the diesel than was normal for the old spark-ignition gas engines.

Caterpillar finally sent out the new engine into practical field use on September 14, 1931. This model, designated the 1C series,

changed the sound and style of Caterpillars—and eventually of all other tractor makers—forever. The prototype, number 1C-1, stayed with Caterpillar in the San Joaquin Valley, California, for testing and development. The four-cylinder engine displaced 1,090ci. In all, 157 of the new series of Diesel Sixties were produced.

Once the problems were sorted out, the advantages—which Henry Kaiser understood earlier and better than most—became clear. In 1932, gasoline for agriculture sold for fourteen to sixteen cents per gallon; whereas, diesel fuel was available at four to seven cents a gallon. In early field tests the Diesel Sixty prototypes performed well under heavy workloads while consuming only four gallons per hour.

Caterpillar carefully monitored who purchased its early diesels. So much was at stake. Breakdowns could be costly in time lost from work, but not only to the new tractor's owner. The effect on Caterpillar's reputation could be deadly. Engineers

◂*This was the twelfth diesel Caterpillar sold, delivered to Mark Weatherford at Fairview Ranch in Arlington, Oregon. The first dozen diesels went to hand-picked locations where Caterpillar engineers could quickly reach them if there was a problem. The diesel was relatively easily started—even in winter storms—using its self-contained two-cylinder gasoline engine. The 25,860lb tractor could pull 11,991lb in low gear.*

▴*Original owner Mark Weatherford's children nicknamed the crawler "Old Tusko" after an elephant at the Portland Zoo. The diesel engine displaced 6.125x9.25in and it produced 54 drawbar and 77.1 brake horsepower at 700rpm. Exhaust from the small two-cylinder gasoline engine was directed toward the diesel cylinder walls and the fuel injectors. Its power was transmitted through a sliding pinion gear that engaged the diesel engine flywheel gear through the action of a hand-lever-operated clutch. Once the diesel engine began to run under its own power, centrifugal force acting on the pinion latches automatically disengaged the pinion.*

◂Weatherford was a tractor-maker's ideal customer. He kept detailed records especially during his first year of use. Working twenty-three hours days, Old Tusko plowed 149.8 acres a day for forty-six days. His total cost was $535.81 for the entire job, averaging 7.8 cents per acre. Caterpillar diesels achieved 13.87/hp-hours per gallon of fuel in tests at University of Nebraska. In 1935 Mark Weatherford sold Old Tusko to Allen Anderson and his family nearby who recently restored the historical tractor.

▸Ford Motor Company, Ltd, of Dagenham, Essex, England, produced both the Major (in the background) and the Dexta. The Major was powered by Ford's in-line four-cylinder 3.94x4.52in engine that produced a maximum of 38.5 belt and 27.7 drawbar horsepower at 1,600rpm. The Major returned fuel economy of 13.93hp-hours per gallon. Both the 1958 Dexta and 1953 Major are owned by Carlton Sather of Northfield, Minnesota.

went along with the first deliveries, staying on site for days to make certain no problems appeared. Mark Weatherford was an early purchaser of one of the diesels. He took delivery of Number 1C-12 early in March 1932 at his ranch in Oregon. For his own curiosity, he kept detailed records of his first work with the tractor. Between March 4 and April 27, 1932, Weatherford plowed 6,880 acres, averaging 149.8 acres per twenty-three-hour day. Even with the Oliver twelve-bottom 16in plow, Weatherford's diesel consumed 5,440 gallons of 7 1/2-cent-per-gallon diesel fuel. Adding in all the other costs of lubricating and transmission oils and one $2.99 repair, it cost him $535.81 for the entire job, or barely eight cents per acre. Weatherford estimated that the 1C-12 saved him $600 in fuel costs alone over his previous year with his 60hp gasoline tractor.

As the success of Caterpillar's diesels became known, Art Rosen underwent celebrity status similar to that of Rudolf Diesel himself. Rosen delivered more than a dozen research papers between 1932 and 1935. Most significant to technical listeners and farmers alike were the performance characteristics of the diesel.

Rosen emphasized the engine's lugging ability. Its "rated load" was usually well below its peak load capacity. This gave operators a wide margin of power for a tough situation. Its torque curve most clearly exemplified the difference between gasoline engines and the diesel: reducing engine speed by 5 percent increased torque by 10 percent in the diesel. But this would improve gas engine torque by only 2 1/2 percent. On a practical basis, this meant that the diesel had superior ability to pull itself out of difficulty without greatly reducing engine speed to increase torque.

Of the first dozen Caterpillar diesels, two went to Hawaii to Theo H. Davies for the cane plantations. Art Rosen went with them. There, he soon learned that the heavier diesel crawler

▶ *The Oliver company dated back to 1868 when Scotsman James Oliver patented the chilled plow, so called because during the process of casting the iron, the outer surface is cooled more rapidly than the center. This gave the plow a harder finish. Oliver opened his shops in South Bend, Indiana in 1853 and by 1900 was widely known for other implements as well. In 1929, Oliver merged with Hart-Parr and several other companies to create Oliver Farm Equipment Corporation. Tractors were called Oliver Hart-Parr into the late thirties.*

◀ *Power-adjustable rear wheels were an option popular with all manufacturers; Oliver sold theirs for $205. This 1954 diesel row-crop Super 66 sold for less than $3,300. Power steering was optional at $198. Both single and dual front row crop configurations were offered in addition to an adjustable wide front axle.*

with no more horsepower than the gas model was actually less effective on the hillsides. To satisfy the operators, he had to open up the smoke screw (allowing more fuel into the mix). This was a trick quickly learned by many other engineers and mechanics after Rosen.

Another nine of the first two dozen went to Eygen-Bilsen, Belgium, in the spring of 1932. The job there was excavating the Albert Canal. The diesel seemed like the perfect solution to farmers', and even contractors', needs. But selling the diesel engine was difficult. With the money involved, no farmer or builder wanted to be first—or even early, for that matter—to jump on the trend.

▼ *The Oliver diesel was a four-cylinder vertical in-line engine with 3.31x3.75in bore and stroke for total capacity of 129.3ci. Nebraska tests yielded 17.7 drawbar and a maximum of 25.0 belt pulley horsepower at 1,600rpm. This tractor was restored and owned by the late Tiny Blom of Manilla, Iowa.*

◀ *Oliver introduced its Fleetline series in 1948, standardizing appearance and a number of parts in the replacements for the 60, 70, and 80 series, known as the 66, 77, and 88. The Super series followed in 1954, with 55, 66, 77, and 88 model tractors offering independent disk brakes, independent PTO, and an onboard live hydraulic system called Hydra-Lectric.*

The sales grew slowly. Caterpillar sold more diesels in 1936 than it had in the previous four years combined. Production was up to 1,000 diesel engines per month. The goal was set for 1,500 per month in 1937. Then came trouble.

Engine problems were occurring everywhere in the world. Piston rings stuck. Cylinder walls scored. Main bearings burnt up.

Caterpillar had about 3,000 diesels in the field by this time. Art Rosen observed that engines running in the western United States on West Coast oils—oils with a paraffin base rather than the asphaltic-based oils from the eastern oil fields—didn't have as much of a problem with rings sticking. It took a cooperative effort between Standard Oil of California and Caterpillar to solve the problem. G. B. Neely was the engineer who worked with Rosen to develop the first detergent oil. Standard called it "Delo." Caterpillar's diesels worked again. Within two years, every manufacturer had a diesel in its line-up. By 1960, diesels constituted nearly half of every tractor engine in production. By the 1990s, diesels are nearly all that is available in tractors.

▴Dwight Emstrom restored and owns this 1957 Model 501 Offset Workmaster Diesel. This tractor is fitted with a single-row cultivator. The offset is 8in to the operator's left of the tractor center line, for better visibility while using implements.

◂This series of tractors was manufactured from 1955 through 1962. It could be purchased with either a four- or five-speed transmission, with or without power take-off or hydraulic lift. It was also available with Ford's Select-O-Shift transmission that used the equivalent of ten gears to keep engine speed in its most efficient power ranges.

▸Ford's engine in the 501 was its own in-line four-cylinder with 3.56x3.60in bore and stroke for total capacity of 144ci. Compression ratio was 16.0:1 (where as the gas version was 7.5:1 and 8.84:1 for LPG). Ford's horsepower figures for the diesel version were 29.3 drawbar horsepower. Ford used the same 144ci engine in the 501, 601, and 701 series tractors.

▲The Ford Dexta was equipped with a six forward-gear transmission that provided a top road speed of 17.3mph. The diesel Dexta weighed 3,393lb, but when ballasted up to nearly 6,000lb gross weight, the three-cylinder diesel was capable of pulling 4,362lb in first gear in tests at University of Nebraska.

▶Ford Motor Co., Ltd. of England, introduced the diesel-engined New Major model at a show in London in 1951 with deliveries starting in 1952. A 16:1 compression ratio was used. This 1958 Dexta model followed the New Major as a smaller companion in the mid-fifties. Ford in the U.S. introduced the diesel for 1958 models.

▲ As early as 1948, Ford of England had offered the Perkins P6 diesel as an option to its Fordson Model E27N tractor. With fuel costs so much higher in England and Europe than in the United States, interest in economical diesel fuel led to healthy sales. But the Perkins was a stopgap effort. Factory engineers had began to develop a Ford diesel engine in 1944.

◄ The Dexta used Ford's three-cylinder 3.5x5.0in vertical engine. Total capacity was 144ci. It was tested at Nebraska in March 1959 and produced 23.0 drawbar horsepower and 31.4 maximum brake horsepower through the PTO at 2,000rpm. This English Ford is owned by Palmer Fossum of Northfield, Minnesota.

▲ Frank Bettencourt's 1948 narrow-gauge D4 was first used in beet fields in California's Santa Clara Valley. It is fitted up to a John Deere-Killifer Model 7MKO-02 disk. Bettencourt, like many collectors, is gathering the implements used by the machines as well as the tractors themselves. The disk is scheduled for restoration.

▸ University of Nebraska tested a Caterpillar D4 in July 1949. In the tests, the four-cylinder 4.5x5.5in bore and stroke in-line diesel produced 33.0 drawbar and 46.5 belt pulley horsepower at 1,400rpm. The tractor weighed 11,175lb and in low gear was able to pull 9,555lb. The five-speed transmission offered a top speed of 5.4mph.

▲*Bettencourt's D4 was fitted with an aftermarket live hydraulic system produced by Be-Ge Manufacturing Co. in Gilroy, California. By shifting the lever rising up behind the seat, the operator could raise or forcefully lower hydraulically assisted implements such as the Killifer disk.*

▶*It is an uncommon story of competitors helping in adversity. Massey-Ferguson's large tractor factory in Detroit suffered a serious fire in 1961, destroying its capacity to produce tractors. Negotiations with Oliver led to license and manufacture of its Model 990GM as the Massey Model 98 through 1962. This unusual 1962 machine belonged to the late Tiny Blom of Manilla, Iowa.*

▲ *Power for the Model 98 came from a General Motors 3-71 diesel three-cylinder in-line vertical engine fitted with a GMC supercharger. Bore and stroke was 4.25x5.0in for total capacity of 213ci. At 1,800rpm, the drawbar horsepower was rated at 61.6 and belt pulley output was 84.1hp. The large exhaust pipe, coupled with the supercharger howl, created a marvelous roar.*

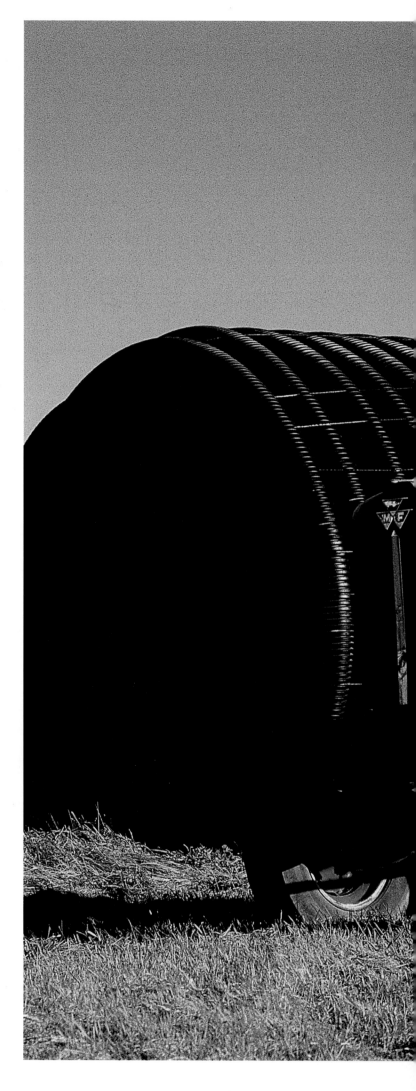

▶ *Iowa farmer and tractor collector Donald "Tiny" Blom specialized in collecting Oliver and Hart-Parr tractors. This Massey-Ferguson, with its Oliver parentage, was one of the more significant pieces of his agricultural machinery collection. An Oliver and Hart-Parr farmer for many years, he passed away in early 1995.*

Chapter Five

A Rubber Revelation

Firestone and Allis-Chalmers Tire the Industry

Agricultural history might suggest that persistence and nonconformity were daily marching orders to most tractor inventors. Such was the case with many managers as well. General Otto H. Falk was brought into Allis-Chalmers on April 16, 1913, to operate the company following its second courtship with bankruptcy in a decade. Falk was a retired Wisconsin National Guard brigadier who was neither a businessman nor a farmer. But lacking these two disciplines meant his imagination was unfettered.

His ideas were innovative; for example, his first tractor, a 10 to 18hp rated tricycle, was built on a one-piece, heat-treated steel frame. Allis-Chalmers' promotional literature boasted that "it has no rivets to work loose," that it "will not sag under the heaviest strains." But its single front wheel was offset; this made steering difficult under those same heavy loads. Falk's nonconformist designs kept early Allis-Chalmers tractors from being accepted or respected by the marketplace.

The second generation Falk tractor was the 6-12. It was conceived as a full-system machine that drove two large wheels far out front, beneath its engine. The implement-of-choice served as the rear-wheel assembly. That one too closely conformed to Moline Plow Company's Universal, and this prompted a note from Moline about patent infringement. By August 1923, Allis had dropped the 6-12. But whether it was discontinued because of Moline Plow's insistence or simply because of farmer disinterest is uncertain.

Ironically, Falk had succeeded back in December 1919. Allis-Chalmers brought to the market a stylish tractor with a 15 to 30hp rating. Its smart appearance suggested current automobile designs. Seen in profile, a continuous line ran from the radiator back to the fenders over the rear drive

▲*The Hercules Gas Engine Company of Evansville, Indiana, produced one-to-fourteen horsepower single-cylinder stationary engines for farm power from 1912 until 1934. Hercules Motors Corporation of Canton, Ohio, began producing four- and six-cylinder tractor and truck engines in 1915. It introduced a four-cylinder diesel in 1932. Hercules was owned by Hupp and then White Motor Corporation from 1961 through 1976.*

▶*Eagle Manufacturing Company was founded in 1888 in Appleton, Wisconsin. Beginning in 1899, they produced gasoline engines and in 1910 they made large, powerful tractors using their new two-cylinder and four-cylinder engines, followed in 1913 by a smaller model. Two-cylinder machines strongly resembling Waterloo Boy tractors remained in production until 1930 when Eagle began buying six-cylinder engines outside.*

wheels and circled down nearly to the ground. Allis-Chalmers reclassified it in February 1920 as the 18-30.

But an accident of bad timing slowed the 18-30 to a complete halt within a couple of years of introduction. The economy had dug into a depression following World War I. Henry Ford had reacted to the sales slow down of his new Fordson by cutting its price by more than half. Those manufacturers who could afford to play along with Ford dropped their prices, too. But many of them who thought they could afford to price their tractors competitively—and many of those who knew they couldn't—went under. Allis-Chalmers sold only 235 of the Model 18-30 in 1920.

Allis introduced a similar but less muscular tractor, the 12-20, in 1921, but it suffered badly from the depression and the tractor price wars, selling only 1,705 up to 1928. Yet business resumed when the economy loosened up. General Falk had taken a beating during the first of his two decades with Allis-Chalmers. He force-fed tractors to a company onto which he had been forced. He had made mistakes in design and engineering, and those had cost Allis money, but he saw the future, and he knew that tractors would be there. Allis must have tractors to survive. So he persevered. By 1926, his favorite department was well-established. Tractor production grew twenty times as great in 1928 as it had been in 1926, with total sales nearing 16,000.

In 1926, Falk hired a bright talent, Harry C. Merritt. By all accounts, Merritt was a progressive innovator. He looked on the Allis-Chalmers' tractors from a similar perspective to Falk's—the view of the future. As Falk's tractor department manager, he was in a position to move the products; and with the department being clearly Falk's favorite son, Merritt received encouragement as well.

▲ Eagle used a Hercules Model QXB-5 in-line six-cylinder engine of 3.25x4.125in bore and stroke. Eagle rated it at 18.0 drawbar and 28.0 belt pulley horsepower at 1,575 rpm. A Zenith carburetor and a Splitdorf HT magneto were standard equipment. The Model 6C weighed 3,250lb, and measured 69in tall, 66in wide, and 130in overall.

▶ Eagle introduced its 6C, the Utility, in 1938. Four gears allowed top road speed of 13mph. Auto-Lite electrics provided starting and lights. This is tractor No. 2,494, and Eagle's numbers stop at 2,500. By 1939, Eagle was selling off inventory, and the company disappeared by 1940. This tractor was restored and is owned by Walter and Bruce Keller of Kaukauna, Wisconsin.

The new Allis-Chalmers Model 20-35 was introduced in 1929, three years after Merritt's arrival. This was a slightly improved version of its predecessor, the 18-30. Also known as the Model L, it was available for only two years. Meanwhile, Allis prepared a new tractor for an outside marketing firm, United Tractor and Equipment (UT&E) of Chicago.

This was to be a three-plow rated machine, using a four-cylinder Continental engine. When UT&E failed—despite seemingly adequate outside support and backing—Allis continued to produce its tractor, marketing it as the Model U.

In the spring of 1929, Merritt traveled west, visiting California in May. He was startled and impressed by the wild poppies that covered the hillsides like bright blankets. Their brilliant orange color struck him.

Returning to Milwaukee, he faced the somber green tractors produced by his department. Allis-Chalmers tractors were invisible in the green fields across North America. He felt his products—improving with every successive model—should stand out against the green of growing crops, and against the other makers' somber colors. "Persian" orange most closely matched the wild meadows of California. It wasn't long before Merritt's orange tractors dotted the fields.

Allis-Chalmers' tractor marketing efforts were given a huge boost with the acquisition in 1931 of the Advance-Rumely Thresher Co. This purchase accomplished what General Falk had not succeeded at doing on his own. Overnight, Advance-Rumely provided Allis-Chalmers with a well-established dealer organization that sold a highly respected line of tractors and implements. Allis' Model U remained in production through 1944, and Allis built more than 10,000 of them. Continental engines powered early Model Us until Allis-Chalmers' own UM four-cylinder engine appeared in 1933. This tractor was offered in a

▲ Production of the Row Crop tricycle began in March 1930 and was up to twenty units a day by April. The single-front wheel was replaced with a dual-front beginning in 1932. In 1937, this same model became the Oliver 80, on rubber tires. This original, unrestored 1931 machine was owned by the late Tiny Blom of Manilla, Iowa.

◄ The Row Crop sold new for $985. Oliver inaugurated its advertising campaign promoting "Power on Tiptoe" with this model. Its narrow rims were fitted with solid iron grousers. The point of these wheels—and of Oliver's advertising—was to emphasize the limited soil compaction caused by these rear, load-bearing wheels.

variety of styles, including an "Ind-U-strial" model, a crawler, a railroad yard switcher built by Brookville Locomotive, and a row-crop version.

But it was not only its longevity, its adaptability, or even its new color for which Allis-Chalmer's Model U is most famous. It was for Harry Merritt's friendship with Harvey Firestone.

In his 1951 history, *The Firestone Story*, Alfred Lief categorized Firestone's relationship with Allis as one of the Firestone Company's "self-helps to recovery" from the financial depression following the war. It was that at the very least.

Firestone's family farm, the Homestead in Columbiana, Ohio, about fifteen miles south of Youngstown, ran on steel and was operated with Allis-Chalmers tractors. Harry Merritt had of-

▲ The Hart-Parr in-line four-cylinder engine measured 4.125x5.25in bore and stroke. At 700rpm, it produced 18.0 drawbar horsepower and a peak of 29.7 belt pulley horsepower in its tests in April 1930 at University of Nebraska. Oliver fitted a Bosch U4 magneto and Ensign Model K carburetor which was protected behind a large Donaldson air cleaner. The tractor weighed 4,650lb.

▲Beginning in 1930, Allis-Chalmers engineers field tested pre-production models of its new Model U fitted with rubber tires on a farm near Waukesha, Wisconsin, belonging to Albert Schroeder. The tests were very successful. Nebraska's board of governors heard about the Allis-Chalmers/Firestone development and invited Allis-Chalmers to bring a rubber-tired Model U for a test, waiving the usual $500 test fee.

▶Allis-Chalmers designed the tractor for the United Farmers Cooperative who were then unable to put the machine into production. An Allis-Chalmers United was tested at Nebraska in November 1929 on steel. Using the Continental four-cylinder 4.25x5.0in engine, 19.3 drawbar and 35.0 pulley horsepower were recorded. The 4,821lb tractor pulled 3,679lb in low gear.

fered Firestone one of the company's new Model U tractors. Firestone, long convinced of the value of pneumatic tires for cars and trucks on the road, wondered about their application to the farm. Surely, for the more than a million farm tractors in use on steel wheels, the same economy of operation, the same operator comfort, the same reduction in vibration would apply—if the problem of traction could be solved.

Firestone understood the conditions. The ground and crop clearance required of a tractor in all farm operations demanded that the tire have a large diameter and a wide cross section for contact with the soil. Yet it still had to fit within the furrow. Contact with the soil had to be firm so the tractor would pull; however, too much air pressure would pack the soil too hard. Yet the grip could not be so great, or the pressure so little, that the tire would creep on the rim.

Firestone's engineers tried flat truck rims first. Airplane tires were mounted on them, but they rim-crept. The answer was a drop-center rim with a tight bead fit. So much for creepage. As for traction, the tread had to bite into earth, sand, wet clay, or sod. However, if they were cut too deep or had insufficient support, the treads could bend under the load. The engineers modified a chevron pattern, inventing a connected bar design, the continuation of one side of the chevron to the bar above it. With Firestone himself at the wheel, the engineers settled on 12 to 15psi for tire inflation.

Firestone's farm was completely re-tired; each tractor and implement—thirty-two in all—was changed over to his new "Ground Grip" pneumatic tire. Merritt changed the specifications on his Model U. Inflatable rubber was offered as an option. To promote some of the virtues of the new tires that were introduced for general sale in January 1935, Firestone and Merritt

▲ *The tire paint is fading but the history is clear. This is a replica of the first tractor tested on rubber tires at the University of Nebraska. It was fitted with 6.50x20 rayon Firestone Transport Heavy-Duty Truck-Bus eight-ply tires on the front and 11.25x24in airplane tires on the rear. French & Hecht manufactured the rims. This replica belongs to the University of Nebraska tractor test museum and it is photographed on the concrete test track that rubber tires eventually made necessary.*

▶ *Five years later, The Model U was tested on rubber and again on steel. While Allis-Chalmers had begun making its own engine, test results with the same tractor on steel and on rubber revealed dramatic benefits to rubber tires. A four-hour fuel economy run produced 6.14 horsepower hours per gallon with 16.9 drawbar horsepower on steel, and 8.57 at 22.7hp on rubber. None of that even began to demonstrate the improvement in operator comfort.*

coaxed Indianapolis racer Barney Oldfield out of retirement. They put him back on the county fair race circuit, on a rubber-shod Model U. *Implement & Tractor* magazine reported that countless events were run. In five lap races against local Allis owners and salesmen, Oldfield let the amateurs lead for the first four laps, then he would sprint past them, showing off the ultimate speed and traction capabilities of the Allis-Chalmers rubber-tired combination. Merritt fitted especially high "road" gears to the racers; on one top speed run, Oldfield set a record of better than 64mph with a specially modified U and Firestone's tires.

The brightly colored rubber-shod tractors garnered a lot of attention, even as farmers acknowledged the uselessness of 64mph machines. The Model U became the first tractor in North America available with rubber. When Merritt introduced Allis' second-generation row-crop, the WC in 1934, it was the first tractor to be designed with inflatable rubber tires specified as standard equipment. Firestone's Super Grip Type R tire improved the connected bar design. It provided greater depth, narrower width, and wider spacing between bars. These entered the soil more effectively and cleaned themselves on exit.

Lester Larsen, head of the tractor testing program at the University of Nebraska for thirty years, recalled that Allis-Chalmers had a Model U on Firestone rubber tires testing for a full year on Albert Schroeder's farm in Wisconsin. At the beginning of 1934, there were only three tests scheduled, a Farmall F-12, a John Deere Model A, and Allis-Chalmers' new Model WC. The University of Nebraska Tractor Test Board of Governors had heard about the tire test in Wisconsin and about work on a Case Model C with Goodyear. The board extended an invitation to IHC, Deere, and Allis-

◂The tiny 1934 Plymouth measured 49in tall (60in to the top of the steering wheel), 48in wide, and 100in long overall. Its wheelbase was 61in, its non-adjustable track width was 38in, and it ran 5.50x16 front tires and 9.5x24 rears. This unusual one-plow tractor is owned by Paul Brecheisen of Helena, Ohio.

▴Built by Fate-Root-Heath (FRH) Company of Plymouth, Ohio, the Plymouth tractor was named after the company's home town. FRH used a Hercules in-line four-cylinder Model IXA engine. In 1935, FRH renamed their tractors the Silver King line that continued in production until the late fifties.

Chalmers to bring their tractors not only on steel but also to bring one—at no extra charge—on the rubber tires. Tests at that time cost the manufacturer $500. Only Allis-Chalmers accepted. During the test, the WC on steel produced 14.4 maximum drawbar horsepower. While on rubber, the same tractor produced 19.6hp and returned a 25 percent increase in fuel economy.

But people are often cautious and sometimes cruel toward new ideas. Where a farmer appeared with rubber tires on his tractor working the field, neighbors stood at the fence and called him "sissy," looking down their noses at the gentle ride. To placate the conservative farmers, steel wheels were still optional, up through the 1940 introduction of the WF standard tread version of the WC. Of course, World War II returned some of the U, WC, and WF models to steel out of necessity. But once the war ended, steel wheels were no longer available on these tractors. Farmers came to understand that while comfort may be sissified, the improved economy and performance definitely were not.

▲The Hercules IXA engine measured 3.0x4.0in bore and stroke. The Plymouth used a Zenith 0.875in diameter carburetor as well as a Fairbanks-Morse magneto (although an International Harvester Corporation magneto has replaced the original Fairbanks unit on this version.) Horsepower was quoted as 10 on the drawbar and 20 on the belt pulley at 1,400rpm.

◀Fate-Root-Heath used a four-speed transmission offering a top road speed of 25mph. FRH built 325 Plymouths before renaming their tractors the Silver King. This was the 39th model manufactured. Plymouths were offered on either steel wheels or pneumatic rubber tires.

▲This UC used Allis-Chalmer's own 4.375x5.0in bore and stroke in-line four-cylinder Model UM engine. It recorded maximum drawbar output at 18.9hp at 1,200rpm. Belt pulley power was 33.5hp. Rubber-tired versions could hit 10mph in fourth gear.

◄When the University of Nebraska tested the Allis-Chalmers Model U on both steel wheels as well as rubber tires, it also tested a new Model UC. Fuel economy was 5.35 horsepower hours per gallon on steel while it jumped to 8.96 on Firestone rubber.

▲One significant feature of Allis-Chalmers tractors was the Snap-Coupler quick-attachment system for implements, which worked especially well with cultivators and similar attachments. Mounted under the tractor between the axles, these reportedly could be fitted or removed in less than five minutes for either operation. This 1937 model is owned by Conrad Schoessler of Westside, Iowa.

▶Allis-Chalmers introduced the Model UC in 1930, and it competed head-to-head with the Farmall Regular. The UC remained in production from 1930—using a Continental engine—through the 1934 introduction of their own Model UM engine until production was ended in 1941.

▲ International Harvester manufactured its own in-line four-cylinder engines. This one measured 4.25x5.0in bore and stroke. In its Nebraska test in October 1931, it produced 24.9hp on drawbar at 1,150rpm, and 32.8 belt pulley horsepower.

◄ The Model F-30 high clearance offered 27in of ground clearance below the drawbar. It stood 99in tall (81in to the top of the air cleaner) and 84in wide as well as 145in overall. This 1937 model is owned by Bob Pollock of Denison, Iowa.

▶ IHC used its own F4 magneto as well as a Zenith carburetor. The high clearance versions were designed specifically for use in sugar cane fields. This tractor sold new for around $1,225. Front tires were 8.0x21, rears were 13.6x44.

▶▶ The Farmall-30, or F-30, was introduced in late 1931, and it remained in production until 1939. Its 94in wheelbase allowed a 17ft-4in turning circle. More than 28,000 Model F-30 tractors were produced.

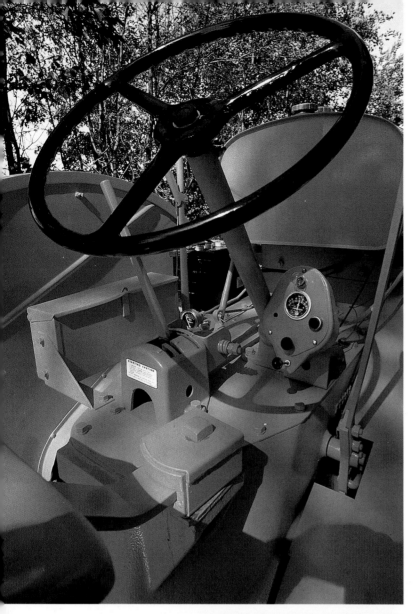

◄ The Model G was introduced in 1940 as Minneapolis-Moline's strongest, heaviest tractor. For the 1942 model year, slight changes were made which allowed it to be renamed the GTA, as it continued to be called through 1946. On steel wheels it weighed around 7,100lb. It used a four-speed transmission with a top speed of 6mph whereas the rubber-tired versions had a fifth gear capable of 13.8mph.

► Due to U.S. government needs for rubber during World War II, the few tractors that were delivered during the War were only available on steel. These steel wheels measured 6.0x37in front and 12.0x50in rears with 5in grousers. This tractor is owned by Walter and Bruce Keller of Kaukauna, Wisconsin. It is a 1946 Model GTA.

▼ Minneapolis-Moline built its own in-line vertical four-cylinder engine for the G. This had bore and stroke of 4.625x6.0in. and total capacity of 403.2ci. A Marvel-Schebler carburetor was used along with a Delco generator and electric system. No tests were performed at University of Nebraska during the War but in 1950, a G on rubber tires was tested, producing 39.2 drawbar and 55.9 belt pulley horsepower at 1,100rpm.

▲ For an additional $55, buyers received rear fenders, the operator's platform, and the swinging drawbar. The stakes that protrude from the bottom of the rear fenders scraped grass and dirt from between the steel wheel grousers. The tractor was also available from as early as 1929 on hard rubber tires for industrial applications. Inflatable rubber came in 1934, at a price of $1,405 compared to $1,175 for the tractor on all steel wheels.

◄ Connie Schoessler's 1927 Model L Case hugs the hillside on his Westside, Iowa, farm. Prior to the introduction of the Model L, many of Case's gasoline or kerosene-fueled tractors had their engines mounted transversely, or "cross-mounted". This model sold new for $1,295.

◂The Case weighed 5,307lb and in first gear it pulled 4,555lb. The company offered the tractor with a three-speed transmission up until 1937 when four-speeds could be ordered on industrial and agricultural models. Top speed for the agricultural version on steel was 4mph; for either version on rubber, top speed was not quite 12mph.

▴Tested in Nebraska in 1929, the Model L did better than its advertised 26-40 rating. The four-cylinder 4.625x6.0in engine produced 30.0 drawbar and 44.0 belt pulley horsepower at 1,100rpm. Case fitted a Kingston L-3 carburetor and a Robert Bosch FU4 magneto.

Chapter Six

Fuel Evolution

FRACTIONS ARE NOT ALWAYS ARITHMETIC

LIQUEFIED PETROLEUM GAS—LPG—WAS just a waste gas. Before its commercial value was recognized, it had nicknames: "wet gas," "greased air," "fizz gas," and "overhead reflux." But once a market was created for it, it was proudly, and profitably, sold as Gasol and Gasolite, Pyrofax and Pyrogene, Readygas and Flamo, Skelgas, Shellane, and Bu-Gas. Generically, it was known as propane and butane.

It had been "discovered" by Dr. William O. Snelling, a thirty-one-year-old consulting chemist on the staff of the U.S. Geological Survey working in the explosives laboratory in Washington. Snelling succeeded in liquefying natural gas and storing it in a thick glass bowl. He lit the vapor above the liquid and used this flaming vapor to light his office for months before ever saying a word about it to anyone. He described the gas in rather technical terms to William Altdorfer, a freelance writer for the *Indianapolis Star*.

"The gas is prepared from 'heavy' natural gas, particularly waste gas which accumulates in the pipes of oil wells," Altdorfer learned in 1912 from his interview with Snelling. "The natural gas is compressed and cooled and the heavy fractions which condense are separated. The lighter fractions are next condensed and are forced under pressure into a vessel called the 'rectifier' where they come in contact with coils of super-heated steam and are completely vaporized. The gases then pass in succession through a series of coils, each heated to a lower temperature than the preceding one, and these coils separate the gas into a series of products…The higher compounds of the paraffin series of hydrocarbons—to which the chemical names of 'ethane,' 'propane' and 'butane' have been given—are liquefied."

In chemical terms, "fractioning" refers

▲*Close to rocket science, the operator's view is an eyeful. A fuel level overflow valve and fuel gauge are just ahead of the standard 971 instrument panel. Below the panel is the controller for the Select-O-Speed transmission that indicates engine speeds at 1,200, 1,750 and 2,200rpm matched to gears to yield ground speeds from 0.6mph up to 18mph. Dwight Emstrom of Galesburg, Illinois, restored and owns this 1961 tractor.*

▶*The 971 stood 66in tall to the top of the instrument panel, 72in wide, and 133in long overall on its 85in wheelbase. Ready for work with water ballasted tires and LPG equipment it weighed about 5,970lb, roughly 100lb more than the same tractor burning gasoline. Front tires were 5.50-16in duals while rears were 12.0-28s. Ground clearance was 27.5in.*

to separating the various elements of any mixture by distillation or crystallization processes. The success of these processes relies on the fact that each element has a different boiling point or solubility that allows its separation from other elements.

If this entire operation sounds much like the working procedures of steam engines or even early gasoline, coal dust, and diesel engines, there is some similarity. A vapor, gas in this case, is cooled and condensed to liquid, whereupon it is heated and expanded into vapor form again, after which it is once again condensed. But this last condensation is done extremely slowly and deliberately. This will separate the various flammable elements from the original liquid or its vapor.

As early as 1860, Julius Pintsch, a German, developed a compressed gas for high temperature oil processes. But it was fifty years later, in 1901, that another German chemist, Hermann Blau, invented a process to liquefy petroleum gas. Patented in the United States in 1904, Blau gas had a foul aroma, but it was used as cooking and lighting fuel. Blau's patent covered bottling it for widespread distribution. It could be sold in shops like so many food stuffs.

Between 1905 and 1910, several LPG pioneers established businesses to deal with separating, storing, and marketing the wet, greasy gases. Frank Peterson of Mercer, Pennsylvania, extracted gas from anthracite coal in 1905, which he then used as fuel for his own small engines. In 1909, Andy and Chester Kerr formed Riverside Oil Company in Pittsburgh, Pennsylvania, to produce natural gas. In the next two years, Riverside built twenty gas plants and launched the boom. In 1912, more than a thousand new plants opened in western Pennsylvania and Virginia.

Gasoline, a hydrocarbon fuel, had experienced a huge surge in demand at the

◂ *Live power-take-off was provided by use of a double clutch that kept PTO power engaged even when the tractor was stopped. The swinging drawbar was designed to function fully without interfering with either the PTO or the hydraulic three-point hitch system. Rear wheel tread width could be adjusted by engine power in 0.5in increments from 56in to 84in.*

beginning the century, due to the increasing promotion of gas-engine automobiles and the development of the gasoline engine tractor. Refining crude oil or petroleum into gasoline requires a series of distillations of the product. This is sometimes called cracking, a decomposition process by heat and pressure that breaks molecules down to form simpler ones of lower boiling point. Each successive step further controls the quality and purity of the product. Because of the repetition necessary, gasoline always remained more costly than gas-oil distillates, kerosene, and other products obtained from the first distillation, or straight-run, process. Blended gasoline, even more costly yet, could provide specific properties; for example, about ten percent of the mixture may be a low boiling point material that would vaporize without heat on the manifold, as is the case when an engine is started the first time on a cold day.

Robert Clay, the managing editor of *Butane-Propane News*, wrote a fifty-year history of the LPG industry in 1962. His interview with industry founder Andy Kerr was entertaining.

Clay set the scene of the early days in his own, vivid writing style: "The gasoline obtained from these plants was an ornery, dangerous product," he wrote. "It contained a considerable volume of petroleum gases that boiled out of the liquid whether it was in storage, in transit, or in use. A. N. Kerr once rather colorfully and modestly described his side of this situation: 'I deserve no credit for being the first to plunge into the sale of butane and propane. I tried in every way to avoid selling these products. I tried to pass them on to the gasoline trade, but they boiled out of our casing-head gasoline and splashed red ink all over our ledger. These liquid gases made our works at Sistersville so dangerous that we were compelled to build high waste lines with Christmas tree outlets [dump and drainage lines that spread out like the branches of a tree] in order to avoid having an excess of dangerous gas float over the Baltimore & Ohio railroad main line adjoining the works [where it could be ignited by a stray steam engine spark with horrifying results]. I was literally forced to put this gas in its proper place. A daily loss of 1,400 gallons (of vaporized gases) at that time amounted to approximately $150 and no Scotchman could overlook that.'"

But capturing and controlling the vapor was where its orneriness and danger came in. Herman Stukeman, one of Andy Kerr's young engineers, filed a report on the Kerrs' Riverside efforts on Christmas Eve 1910. "While we were condensing the first liquid, our gas line broke and filled the room to a depth of one foot with gas. Apparently the gas did not have enough air mixed with it to burn, except above the one-foot depth. Two helpers in the engine room—who, as helpers will do, were watching the boss repair the leak—had rings burned around their trousers one foot from the floor; otherwise, they were untouched.

"The supervisor [Andy Kerr] being in the fire zone and on his knees, had within the week a new and thinner skin on his face and hands. On retiring from the room, as all did promptly, the gas was observed to be burning from the top, about twelve inches from the floor. No harm was done except to the complexion.

143

◂The engine of the gas and kerosene tractors was the same except for a dual-fuel carburetor. The four-cylinder engine of 3.19x3.75in bore and stroke produced 14.9 drawbar horsepower at a fuel economy of 8.8 horsepower hours per gallon of tractor fuel. The gasoline version produced 17.7 drawbar horsepower at 10.13 horsepower hours per gallon of fuel. The advantage of kerosene was simply that it sold for about half the price of gasoline.

▸The "A" in the model designation 8N-AN stands for "all fuel" or "tractor fuel." Dual filler caps separated one from the other. This dual-fuel Ford belongs to Palmer Fossum of Northfield, Minnesota. In June 1950, an all-fuel Ford 8N and a gasoline 8N were tested at University of Nebraska, revealing significant differences in performance and economy. This is a 1952 version.

◂The LPG tank holds 24gal with a buffer for fuel to expand as it heats up during a routine work day. The gas is fed through a filter, vaporizer, and regulator before reaching the carburetor. The four-cylinder engine measured 3.90in bore and 3.60in stroke for total engine capacity of 172ci. Power output was rated at approximately 43.0 drawbar and 50.0 belt pulley horsepower.

"After our fire experience," Stukeman concluded with understatement, "we had considerable respect for this gas.... (From then on) we used welded containers and gas regulators and reduced the gas to low pressures."

Independently—and in much greater safety—research chemist Snelling had determined through scientific experiments what Andy Kerr had learned painfully: that liquefied petroleum gas was a mixture of propane, butane, pentane, and other separated gas mixture elements.

Still, the dangers were not past. A few months later, one of Kerr's employees at their Sistersville plant was frozen by escaping butane and died. Through fire and ice, each passing horror ex-

▲ *In an indication of the move to alternate fuels, Minneapolis-Moline (M-M) offered this tractor only with diesel or LPG fuel. The six-cylinder 4.25x5.0in bore and stroke engine provided a total capacity of 425ci. At 1,500rpm, Minneapolis-Moline rated the tractor at 82hp on the PTO. Fuel capacity is 41gal of LPG. With its Ellwood mechanical front wheel drive system, it was rated as a nine-plow tractor.*

◀ *The G704 stands 96in tall to the top of the muffler, 76in wide, and 151in long overall on a 85in wheelbase. In LPG configuration, it weighs 8,550lb. Top road speed in 5th gear is 17.1mph. It operated on 9.2x24in 4-ply front tires and 18.4x36 6-ply rears. Power-steering was standard equipment. Bruce Keller operated the 1962 heavy-weight near his home in Kaukauna, Wisconsin.*

panded the body of knowledge. Snelling joined the Kerrs in late 1911, and together they founded American Gasol, the name Chester Kerr had given to the propane fraction a year before. Two years later, they were all bought out by E.W. DeBower, the founder of America's most successful home study programs, LaSalle Extension Course. DeBower had done his homework. He learned a great deal about the economic potential of the gases. He offered the partners $50,000 for their company, and all but Snelling were glad to be out of the risky, if suddenly profitable, business.

Development continued. During World War I, Dr. J. B. Garner of Hope Natural Gas Company in Pittsburgh, Pennsylvania, introduced the word "butane" to consumers. Garner worked to improve its image. Nicknamed "greasy air," the gas had been previously promoted for industrial uses; many chemists and engineers still thought it was too risky for home use, despite the fact that butane had successfully powered an automobile in 1912. But by 1920, the Kerrs, back in the business again, had concluded that the greasy air would "work very well in practically any gasoline-type burner." In 1922, LPG sales totaled 223,000 gallons nationwide. And the Kerr brothers, in from the beginning, sparked LPG's next big flash.

Andy Kerr moved to Long Beach, California, and set up the Imperial Gas Company in 1925. As he explained later to editor Robert Clay, "It was thought that the Imperial Valley [a desert farming area in Southern California between Indio and El Centro] would be a good market. The first set—a revaporizing kind of wet gas outfit—was too expensive, as it consisted of two regulators, one of which used liquid, and it had three tanks...While this plant would use any type of cheap product, it sold for $125." Not cheap in 1925.

When Kerr concluded that this was far too costly for the California farmers, he simplified the package into the first successful tank-vapor system using one tank and one regulator. This was intended for home heating and cooking, and Kerr was assisted by both Standard Oil Company of California and by Shell Oil Company, which was poised to introduce its own propane-propylene mixture, Shellane. Standard's Long Beach-based Lomita Oil Company soon brought out Readygas, which it test marketed and introduced as Flamo. And Phillips Petroleum, at that time the largest natural gas producer, began to examine the challenges, logistics, and value in transporting liquefied petroleum gases by rail to distributors around the United States. As home owners in rural areas began to accept butane and propane for home heating and cooking uses, the manufacturers continued working out the kinks in the hardware.

Clay described an ongoing struggle: "It was not uncommon for domestic sets to lose many pounds of fuel each day. Because LPG not only operated under higher pressure, but also had pronounced solvent qualities, the loose fits, rubber compounds, and grease packing that worked with natural gas would not work very long with LPG." Yet even as the chemists and researchers, tinkerers, and inventors struggled with these problems, other individuals sought out new uses for the fuels.

Standard Oil had established Lomita Oil in 1923, pumping gas and fractioning it from its pumps and fields on Signal Hill at the north end of the city of Long Beach. Sometime in 1928, as Lomita began to promote the commercial uses of propane and butane, Charles McCartney, a Stanford University engineering dropout, walked in the front door with an idea. McCartney believed that either of the two gases could be used as motor fuels. He asked Lomita to put him in business distributing gas for this virtually untapped market. He quickly founded Petrolane and set out selling butane as an engine fuel to farmers from El Centro to Bakersfield.

At that time, in 1928, Los Angeles-based George Holzapfel—among others—had begun work to develop LPG carburetors. A year later, Shell had a truck in its fleet testing the fuel daily. In 1930, due in part to McCartney's vision, Standard of California announced that 127 of its salesmen would be driving its Flamo-powered trucks. As Clay pointed out, Flamo was propane, and the large Standard Oil program produced a huge butane by-product surplus.

This was, of course, a time of boom and bust. McCartney was selling a conversion that cost $150 to $250 for a farm tractor. But once it was converted, the tractor ran on fuel that cost four or five cents a gallon compared to eighteen cents for gasoline. In addition, butane burned so much cleaner that routine maintenance such as oil changes could be stretched to nearly double the engine running time. Furthermore, butane, as a fuel, simply produced more power than gasoline. Other inventors and manufacturers joined as outside suppliers to the volatile market. Still others enlisted as inventors-turned-disciples.

Roy Hansen, a licensed mechanical and civil engineer in Lomita, was in business manufacturing carburetors and tanks in 1933. He formed an alliance with Petrolane, and Hansen and McCartney began traveling north to Stockton and Sacramento and beyond to find new customers and new distributors.

Converting tractors—in the early days these were largely Caterpillars—was not extremely difficult, though it was involved and it did require some internal engine modification as well as external part replacement. The gasoline tank was replaced with a sixty-gallon butane tank. These tanks were much heavier and contained a safety valve set at about 180lb pressure. The regulator was mounted on the carburetor itself. Sometimes it was called a vaporizer.

The condensed liquid fuel was ice cold. It had to be heated to make it vaporize. To start a Caterpillar Model Sixty, the motor had to run on vapor out of the top of the tank for about ten minutes. It took about this long to warm up the motor before the coolant was warm enough to heat the fuel.

But this lean mixture would burn the valves if the engine ran too long just on the vapor. So after the motor reached operating temperature, the vapor valve was turned off, and the liquid valve was opened. The regulator would vaporize the fuel before it got to the carburetor.

Converting an engine to run on butane required increased compression in addition to the carburetor modification and the addition of the regulator. For this reason, most conversions were sold most easily and accomplished most effectively when the tractor was down for an overhaul. Gasoline at that time rated 61 octane; butane rated 93 and propane was 125. The higher compression in the engine was necessary to avoid "pinging," a pre-ignition detonation of the fuel that burnt valves and could burn holes in pistons. Under normal practices, the converters put in high-rise pistons on Caterpillar Thirties, and they shaved the cylinder heads 3/4in on the Sixties. But farmers have always been tinkerers, and some of those tinkerers fancied themselves as hot-rodders too. "What if" became a question that was never asked by the people who already thought they knew the answers.

Farmers were advised to be careful. Grinding off too much from the heads meant the valves might hit the pistons. Still, farmers experimented, reasoning that if high-rise pistons were good on Model Thirties and shaved heads worked well on the Sixties, at least one farmer believed that doing both would be better. When he started up the engine, it ran for a few dozen revolutions. It made a huge racket, sounding more like a jackhammer than a tractor engine. And then it blew the crankcase studs—those holding the cylinders to the crankcase casting—out of the block.

In 1929, LPG sales had reached nearly 10 million gallons. In 1934, 56 million gallons were sold. Farmers in rice and grain country had adapted their driers to butane. In the late thirties in the Imperial Valley, dirt auto race tracks were so rough that the racers' carburetors would not stay in adjustment. Petrolane collaborated with Ensign Carburetor, which had by this time accumulated extensive experience in making LPG carburetors for dirty, dusty, high-vibration tractor conversions. The Ensign "Special" became the car to beat, and it excited local farmers with its

▲*Minneapolis-Moline advertised that its four-wheel drive system might save owners the expense of buying a crawler. The traction of its system alleviated the need for extra wheel weights. The advantage over crawlers certainly would be the ability to transport itself down the road. The full hydraulic system was available with an optional hydraulic jack that operated out of either pair of rear hydraulic hoses.*

LPG-powered performance. Then, in 1941, Minneapolis-Moline introduced the first factory-produced LPG tractors.

Testing of odorized gas began in the mid-twenties, initiated by the U.S. Bureau of Mines. The bureau found that no single odor was perfect: some people have colds, others can't smell at all, others didn't notice certain odors. Emerson Thomas, a chemical engineer with Phillips Petroleum, began odorizing Phillips' products before the U.S. regulations went into effect. Standard Oil of California tested thiophene, mixing 6.5lb per 10,000 gallons, but found it unsatisfactory. Chemists throughout the industry tested ethyl mercaptan, something the *Guinness Book of World Records* calls the smelliest substance on Earth. This is still in use, mixed at 1lb per 10,000gal. It became the U.S. standard, published in U.S. Bureau of Mines *Pamphlet 58*, in May 1932.

When war made gasoline unobtainable, LPG fell first under the rule of the War Production Board, which didn't limit production but hampered shipments due to the severe railroad tank car shortages. In 1943, the petroleum administrator for war took over LPG, and the Office of Defense Transportation moved LPG off the rails and onto the highways. By the time it all ended, hundreds of thousands of tractors throughout North America were running on LPG.

Chapter Seven

Industrial Design Comes to the Tractor

SUDDENLY, IT'S STYLISH TO BE A FARMER

IN MID-AUGUST 1937, ELMER MCCORMICK of Waterloo, Iowa, arrived unannounced and without an appointment at the fifth floor offices of Henry Dreyfuss Associates at 501 Madison Avenue in New York City. Dreyfuss' secretary, Rita Hart, listened to McCormick in polite amazement. Then she rushed into Dreyfuss' office and blurted out, "There's a man in a straw hat and shirt garters out there who says he is from Waterloo, Iowa and…"

"Where?" Dreyfuss asked. "Never heard of Waterloo, Iowa."

"He says he is from John Deere and he wants to see you about doing some work."

"Who," replied Henry Dreyfuss with growing interest, "is John Deere?"

While Elmer McCormick waited, Dreyfuss and Hart scrambled through the Standard & Poor's *Directory of American Corporations*. And then they opened the door and invited Mr. McCormick in to have a seat and to tell them what they could do for him and for Mr. Deere.

McCormick had traveled 1,100 miles by rail to get to New York. It had given him plenty of time to think of what he would say.

"We'd like," he began, "your help in making our tractors more salable," he concluded. And with that, he'd said his piece.

Henry Dreyfuss was born in 1904 in New York City and had graduated from the Ethical Culture School in 1922. He apprenticed for a year with Norman Bel Geddes, designing theater costumes, scenery, and sets. On his own by 1923, Dreyfuss worked for the next five years overseeing production of weekly stage shows for the Strand Theaters. By the end of 1927, he was burnt out, and he escaped to France. But he returned to New York a year later and began to solicit work as a designer. In 1929, he moved to 580 Fifth Avenue, labeled himself an industrial designer, and began to work to convince manufacturers of their need for his services.

Through the next several years, he devised and formalized his philosophy into what he referred to as a five-point formula. It was based on emphasizing an object's function, that its form should result from its intended use. He paid particular attention to such considerations as the utility and safety of the object, its ease of maintenance, its cost to produce, its sales appeal and, last but not least, its appearance.

The motto, really a treatise, in the office lobby explained the practical application of Dreyfuss' philosophy.

"We bear in mind," it said, "that the object being worked on is going to be ridden in, sat upon, looked at, talked to, activated, operated or in some other way used by the people individually or en masse. When the point of contact between the product and the people becomes a point of friction, then the industrial designer has failed. On the other hand, if people are made safer, more comfortable, more eager to purchase, more efficient—or just plain happier—by contact with the product, then the designer has succeeded."

With that message in front of him, McCormick knew he had come to the right place. After a night in the Waldorf Astoria, McCormick joined Dreyfuss, and they took the morning train headed back west to Waterloo.

What Dreyfuss accomplished has been characterized as merely "a clean up." Perhaps this was the perspective from the designers who would participate in the many great changes to come. But to the farmer, the appearance and the function were so much improved that a word was coined to refer to the effect. Henceforth, tractors—whether "cleaned up" or changed greatly by Henry Dreyfuss or his colleagues, Raymond

▲*Henry Dreyfuss referred to his work on the Model A and B tractors as a "clean up." But it announced that more was to come. The first effort yielded a redesigned grille and sheet metal surrounding the steering gear and radiator, with a new operator's seat and foot platform. Within a couple of years, Dreyfuss and his associates would be involved in complete tractor design beginning at the pre-planning stages.*

▸*Initially, Deere's request to industrial designer Henry Dreyfuss was for help in making the tractors "more salable." The economy had finally turned around after the Depression and advertising and sales people wanted something they could boast was "new and improved" to get the farm family visit to the local agent for a look. These are both 1938 Model BWHs.*

▸▸ *Dreyfuss' first sketches for Deere and Company tractors were dated August 1937, shortly after his first meeting. The tractors he drew were Model Bs that incorporated the steering wheel shafts and gear and the radiator behind a streamlined, powerful-looking steel housing. There are dozens of differences between the two machines, both sold in 1938, both Model Bs.*

◂ *The 1936 Silver King R72 ran on 10.00x24.0in rear tires and a single 6.00x16 front. It measured 67in tall, 78in wide, and 128in long overall, on a 70in wheelbase. It weighed only 2,200lb. It was offered on steel wheels or low-pressure inflatable rubber and was tested both ways at the University of Nebraska.*

Loewy or Walter Dorwin Teague—were forever known as "styled" tractors. Those not bearing the industrial designers' improvements remained "unstyled."

It was a curious word choice. Industrial designers balk at being called "stylists." And the tractors received much more than mere appearance improvements. There were, even from the start, fundamental design changes, and subsequent tractors received engineering changes as part of the Dreyfuss industrial design process for Deere & Co. This was quite similar to work done for International Harvester by Raymond Loewy. But the manufacturers already had engineers and designers on the payroll and some method had to be designated to explain which of the tractors were new and which were old to the board of directors, the branch salesmen, and the farmers. The vague word "styled" did that quite simply.

Before the end of August 1937, Henry Dreyfuss and Associates had prepared more than a half-dozen design studies for new

▴ *Deere & Co. knew its customers and even with its confidence in Dreyfuss' work, Deere hedged its bets. It recognized that as desirable as "new and improved" was to some people, to others, this was all too new. The old didn't need improvement anyway. For several years, Deére continued to sell unstyled machines right alongside the styled ones.*

sheet metal for Deere's Model A and B. A combination of elements from several of the studies became the landmark styled John Deere tractor. These new lines enclosed the steering column and radiator behind a strong-looking grille. But the improvements also affected function, because the narrower radiator cowling and gas tank covering improved visibility forward and down. The instrument panel was redesigned and better organized for the farmer's readability while bouncing through the fields. The back end of the tractor received attention too, as the Dreyfuss designers "cleaned up" its appearance, at the same time making it easier to recognize the different functions of the various fittings.

The tractor seat was an object of serious concern to Dreyfuss and his designers. Farmers sat behind a vertical wheel almost in front of their faces. But with the small rear wheels common in the late twenties and early thirties, every bump over a plowed field was magnified; standing at least allowed the legs to absorb the shock.

So the steering wheel was, logically, set up high. When the operators stood, they could lean against it, putting their hand right on top of it. What was worse, however, was that even with the tractor at rest, the seat didn't fit the average human body.

Dreyfuss once asked Deere & Co. management how they de-

▲ *The 1936 Model R72 used a Hercules type IXB in-line four-cylinder engine of 3.25x4.0in bore and stroke. A Fairbanks-Morse RV4 magneto and a Zenith 1.5in carburetor were standard. At 1,400rpm, the engine produced a maximum of 19.7 belt pulley horsepower and 16.4hp on the drawbar. The four-speed transmission offered 25mph top speed on rubber tires.*

◀ *In the foreground, the 1936 Silver King Model R72 faces off against the 1940 Model 41. The differences are large despite the styling similarities. Both were manufactured by Fate-Root-Heath Co., of Plymouth, Ohio. Both were restored and are owned by Paul Brecheisen of Helena, Ohio.*

▲ *The 1936, shown here, used the Hercules. Later models used a Continental Red Seal type F162 four-cylinder with a Marvel-Schebler carburetor and a Bosch magneto connected to Delco electrics. Top speed was nearly 30mph. Fate-Root-Heath continued to produce the Silver King until 1956.*

◀ *The in-line four-cylinder engine had 3.875x5.25in bore and stroke. At 1,200rpm, the tractor produced 16.3 drawbar and a maximum of 27.2 belt pulley horsepower. It used a Kingston 1.25in carburetor and a Bosch U4 magneto. The tractor weighed about 4,300lb on rubber tires. This 1937 model was restored and is owned by Wes Stoelk of Westside, Iowa.*

▲ *The Challenger was introduced in 1936 as Toronto, Ontario-based Massey-Harris' first row-crop tractor. It was offered on steel or rubber and production ended at the end of 1938. The curved boiler plate sheet beneath the engine served not only as the bottom engine and transmission case but also as the structural frame for the tractor.*

signed their early seat. Dreyfuss had developed his Human Forms, a thorough collection of measurements and dimensions for a fictional Joe and Josephine. Those male and female "models" allowed Dreyfuss to design virtually any action or apparatus to fit any size person. The Deere tractor seat did not fit *any* Joe or Josephine.

Elmer McCormick told Dreyfuss that they used Pete. Deere managers looked around the factory one day to find the fellow with the biggest behind, and they had him sit in plaster. That became the seat size.

Styled Model A and B tractors were introduced to the public for the 1938 season. While these were at first viewed with caution by the same farmers who had vilified the sissies on rubber tires, they were quickly taken up by the branch salesmen who, at last, had something dramatically "new and improved" to sell. When the advertising brochures and service manuals were produced for the new-looking Model A and B, Deere pronounced them "Tomorrow's tractor today."

The risk of selling yesterday's tractor today forced International Harvester, Massey-Harris, Ford, Cockshutt, Oliver, Minneapolis-Moline, Allis-Chalmers, J. I. Case, and many of the independents to jump into the Streamline Age bandwagon with their checkbooks open. Raymond Loewy was hired to do not only IHC's tractors but also its showrooms and even a new, bolder corporate logo. His creation resembled a front view of a row-

▲ The International Harvester Company (IHC) built its own engines, this one being an in-line four-cylinder of 4.25x5.0in bore and stroke. Running at 1,150rpm, the engine produced a maximum 24.8hp on the drawbar and 30.3 belt pulley horsepower. IHC made their own magnetos; this one was the E4A model. The Zenith K5 carburetor was standard equipment.

▶ The F-30 was introduced on steel but, as all manufacturers adopted rubber, the choice was left to the purchaser. Ray Pollock of Vail, Iowa, acquired this tractor in 1948 and, until his recent retirement from active farming, used it every day year round. In more than forty-five years of use, his most major repair was to grind the valves on two separate occasions.

crop tractor with an operator in place.

Industrial design arrived at different times and in different degrees at other manufacturers. In 1928, brothers Joseph, Robert, and Ray Graham acquired the Paige automobile company and began producing striking, sleek automobiles. A decade later, they were manufacturing a general purpose tractor that was sold through Sears Catalog and was as stylish, attractive, and practical as any of their cars.

In some cases, industrial design changed the colors of tractors. In 1938, all of the green tractors carried over from the Wallis designs became red and wrapped in louvered, streamlined sheet metal as part of the Massey-Harris line. Minneapolis-Mo-

▲ International introduced the original Farmall model for sale in 1924, at $950. By late 1931, when the Farmall-30 was introduced, prices had increased only to $1,110. For the model year 1937 and 1938, the price on inflatable rubber tires was $1,225. IHC sold F-30s into 1940, using up inventory on hand after engine production ceased in late 1939. This is the 1938 model.

▶ When Massey-Harris entered the age of "styled" tractors, no company made a more dramatic change in appearance between its 1937 models such as the Challenger and the streamlined Super Twin-Power 101s. The multitude of louvers was a visual cue taken from automobile designs of the late thirties. But the louvers did not offer enough airflow to cool the hard-working engine. Later Massey-Harris tractors had open sides.

◀ In 1920, farms in the United States operated on 25 million horses and mules and 200,000 tractors. It took a war in Europe and then a much different price war back home to change those proportions. The post-war Depression slowed down the conversion through the early thirties. But the introduction of cost-effective, powerful tractors such as the Farmall F-30 turned the tables. By 1940, the horse population was down to about 14 million while nearly 1.6 million tractors now worked the fields. Nearly 250,000 tractors were manufactured in 1938 alone.

line evolved from gray with red trim to prairie gold. On Friday, September 23, 1938, it hosted 12,000 guests at its Farm Equipment Style Show at the Minneapolis Auditorium to show off its stunning new Model U-DLX Comfortractor, one piece of the fall collection of new Visionlined tractors. These machines looked more modern than some automobiles available at the time and, in the case of the U-DLX, they were even meant to replace them. In 1957, Cockshutt introduced its aggressive, purposeful-looking 500-series in two-tone paint schemes, designed by Raymond Loewy's group. These wider, more massive tractors replaced the art-deco stylish delicacy of the nearly all-red 1947 Model 30 that had been designed by architectural designer Charlie Brooks.

Industrial designers are interdisciplinary. Although Norman Bel Geddes, Walter Dorwin Teague, and Raymond Loewy all came from advertising, they and Henry Dreyfuss quickly under-

▲ *The startling styling wasn't the only new development to come out of the Massey-Harris design department. A complete change in color sparked up the appearance of Massey-Harris tractors on the fields of North America. As the continent moved out of the Depression, advertising and sales managers sought to make the tractors more eye-catching even as the engineers worked to make the machines more efficient.*

▸ *This Super 101 sold new in 1938 for just about $1,750 with rubber tires. Massey-Harris had manufacturing facilities in Racine, Wisconsin by this time. A six-cylinder Chrysler type T57-503 truck engine was used. The twin power name referred to using two different engine speeds for power output ratings.*

▲ *The engine measured 3.125x4.375in bore and stroke. Its maximum drawbar power was 31.5hp at 1,500rpm while its peak belt pulley performance was 40.7hp at 1,800rpm, hence the Twin-Power name. A Marvel-Schebler 1in carburetor and an AutoLite electrical system with starter were standard. This machine was restored and is owned by Wes Stoelk of Westside, Iowa.*

◄ *One of the most controversial products of styling was the Comfortractor. Minneapolis-Moline fitted the Model U not only with the enclosed, heated three-plus seat cabin but also with a road gear capable of 40mph. With the Model U-DLX, Minneapolis-Moline hoped farmers would not need an automobile. On Sunday morning, simply hose off the tractor and drive the family to church. Farmers had a built-in radio to keep them company during field work. The side windows rolled down and windshield wipers swept away the rain.*

▲ *The striking 1938 U-DLX was not an unqualified success. Mechanically, it was a reliable tractor. But farmers, accustomed to modesty and humility, looked on it as excessive. Even the justifiable pride in ownership that some owners felt about their Comfortractors put others off. Somehow, it seemed nobler to freeze in the winter or soak in the rain. In the end, barely 125 were sold. It would be years before any other manufacturer would introduce fully-enclosed, heated cabs.*

stood that they needed to be fluent with the languages and techniques of engineering, mechanics, manufacturing, sales, marketing, and metallurgy to make their work most effective.

As Jeffrey Meikle observed in an essay in the 1993 book, *Industrial Design: Reflection Of A Century*, "Dramatic evidence of the economic benefit of designing for mass production came in 1927, when Henry Ford abandoned production of his beloved Model T, a car that transformed American life, and spent $18 million on the newly designed Model A. This 'most expensive art lesson in history,' as someone called it, offered proof to manufacturers that visual appearance had become critically important in marketing ordinary consumer goods. As the business recession of the late twenties expanded into fully fledged economic depression, manufacturers turned to product redesign, at first as a tool for overcoming competition in their own industries, but later as a panacea for restoring the nation's economic health."

Like the electric self-starter replacing the crank, industrial design was seen first by the manufacturers as an easier way to restart farmers' interests in buying new machines. Exactly like the electric self-starter, it became the means for farmers to understand how much easier the tractor was to use, and then to imagine how much more it could do for them.

▴*Oliver used its own in-line six-cylinder engine with 3.5x4.0in bore and stroke. At 1,600rpm, the engine produced 29.1 drawbar horsepower and a maximum of 41.0 belt pulley horsepower. The tractors offered either a four-speed forward/four-speed reverse transmission, or one with six speeds forward. Front tires were 6.00x16 while rears were 14x26.*

◂*Morning sun melted off the hard frost and created a haze in the field behind the late Tiny Blom's 1947 Model 88 Standard. Oliver's designers produced some of the most attractive tractors with the cleanest lines of any of the makes during the era of streamlining and styling. The Model 88 was introduced in late 1947 and remained in production until 1954.*

▲The Standard measured 74in high, 72in wide, and 132in overall. Two transmissions were offered. One provided four speeds forward and four reverse. The other had six speed forward and two in reverse. Top road speed in sixth gear was 11.8mph. Front tires were 6.00x16 while rears were either 13x26 or 14x26.

▶The Standard 88 sold new for $2,603 in gasoline, $3,456 in diesel. Oliver referred to their sheet-metal enclosed streamlined tractors as the "Fleetline Series." Oliver discovered as Massey-Harris had found, that fully enclosed engines suffered from insufficient air flow for cooling. Later models had open sides.

▼*Offset seating and steering kept body width to a minimum on the Model UTC high clearance tractor. Minneapolis-Moline used its own in-line four-cylinder engine with bore and stroke of 4.25x5.0in for total capacity of 283ci. At 1,300rpm, it recorded a maximum of 33.5hp off the drawbar and 37.2hp on the belt pulley. The tractor weighed about 5,750lb.*

▶*The high-clearance model allowed 24in below the drawbar. The tractor stood 105in tall to the top of the exhaust, 81in wide, 134in long overall, and had an 80in wheelbase. Front tires were 7.50x18 while the rears were 14.9x38. Most high ground clearance tractors were destined for work in cotton and sugar cane applications in the southern states and Hawaii.*

▶*Minneapolis-Moline used the term "Visionlined" to describe what Deere & Co. called styled. These new harvest gold tractors were introduced to thousands of customers in a "Style" show in the Minneapolis Auditorium in 1938 as direct competition with Deere, Massey-Harris, and International Harvester. This is the 1946 Model UTC.*

◂◂Case introduced its Model VAC tractors in 1942 and sold them into the early fifties. Rear tread width was adjustable from 48–88in. The 1949 wide-front has the three-point Eagle-Hitch system. These two tractors were restored and are owned by Raynard Schmidt of Vail, Iowa.

▸The Case engine measured 3.25x3.75in bore and stroke and produced 12.5 drawbar and a maximum 17.0 belt pulley horsepower at 1,425rpm during its University of Nebraska tests. A Marvel-Schebler carburetor and AutoLite electrics were standard equipment. The wide-front version weighed slightly more than 3,200lb. Front tires were 6.00x12 on the 1948 single, 5.00x15 on the wide and rears were 11x28 on the wide, 10x28 on the single.

◂The crank screw on the left hitch arm was used to adjust the arms for implement depth or to level them for some harvesting purposes. The top link was removed here because the owner uses the wide-front VAC to tow wagons during corn harvest. Other implements such as cultivators were attached by lever linkage ahead of the rear tires.

▾The 1949 wide-front Model VAC was only one of a variety of VA-tractor configurations that J.I. Case offered. An industrial model VAI was introduced with the entire series in 1942. In 1943, Case offered the first orchard tractors, the VAO, with much of the engine, rear tires, and operators' position enclosed in sheet metal. A single-front and a dual-front row crop model were available and in 1948, a VAH high-clearance model was introduced. For 1952, the VAS provided high-clearance with an offset operators position, the better to cultivate single rows of tomatoes, tobacco, and cotton.

▲The Model R was introduced in 1939 and in 1940 it was offered with a fitted cab after the style of the Model U-DLX Comfortractor. A number of high-clearance versions of these enclosed tractors were modified with baskets alongside the engine and over the front wheels; these were produced for the U.S. Post Office for use as rural mail carriers.

▶Minneapolis-Moline was the first company to bring cabs to a wide audience. Both the U-DLX and the RT used a rear door, which obviously precluded any possibility of utilizing the power-take-off shaft. The tractor measured 83in tall to the top of the rear light, 79in wide, and only 119in long overall. Front tires were Firestone 4.0x15 while rears were 9.0x36.

▲Minneapolis-Moline's in-line four-cylinder 3.625x4.0in engine had a total capacity of 165ci. At 1,500rpm, 18.3hp on the drawbar and 25.9hp on the belt pulley were achieved. Top speed of the non-postal service-modified Rs was 13.2mph. The R sold for $1,500 with cab. This example was restored and is owned by Walter and Bruce Keller of Kaukauna, Wisconsin.

▲The cab interior of the 1951 RT was spartan compared to the U-DLX. Without radio, heater, rear-view mirror, or fold-down seat for a passenger, farmers were more inclined to look at the RT as a proper machine. The seat in the R swiveled for operator ease in entry and exit. The cab itself was 42in from windshield to door, 36in wide, and 55in tall.

▲Rodney Ott's well-used 1954 Super 88 Standard worked nearly every day of its life tending his farm near Hilbert, Wisconsin. As the successor to the Oliver 88, one distinctive visible change was that the side panels around the engine were opened for better air circulation. Oliver offered single and narrow row-crop front axles as well as both adjustable and fixed tread wide fronts.

▸The Super 88 sold for around $3,800 when it was introduced in 1954. The Oliver 256ci six-cylinder engine had bore and stroke of 3.75x4.0in. During its Nebraska tests, it produced 36.8 hp on the drawbar and a maximum of 49.6hp on the belt pulley. The Super 88 would reach 11.75mph in 6th gear at 1,600rpm. It weighed 5,513lb with fuel and operator.

▶*Just as nearly all the other makers had discovered, fully enclosed side panels, even if fully louvered, trapped too much heat. The Model 333 used Massey's four-cylinder 3.69x4.875in engine. It yielded 29.7hp at the drawbar and a maximum of 39.8hp on the belt pulley at 1,500rpm.*

▼*The Model 333 featured a ten-speed forward/two-speed reverse transmission, offering a speed range from 1.5 to 14mph. Row-crop versions ran on 5.50x16in front tires and 12.0x28in rears. The late fifties marked the end of streamlined styling as makers moved toward more massive, aggressive appearances to reflect the increase in power that was coming from the engineers. Wes Stoelk and his son Scott of Westside, Iowa, restored and own this 1956 tractor.*

▲As good as this hitch was, it was about to get better. Massey's Depth-O-Matic hydraulics would quickly become the legendary Ferguson three-point hitch. Power steering was also available on the Model 333. The 333 could also be had in standard and high-clearance models. The tractor was sold only in 1956 and 1957 and retailed for $2,838 for this gasoline row-crop version.

◄In 1953, the sun set on one Canadian manufacturing name and rose on a more international blend. Massey-Harris added the name of Harry Ferguson when it merged with the Irish inventor of the three-point hitch. New tractors under Ferguson's influence didn't appear until 1959, however.

Chapter Eight

Ford, Ferguson and the Three-Point Hitch

The Handshake Heard 'Round the World

"What a waste it is," Henry Ford once remarked, "for a human being to spend hours and days behind a slow moving team of horses." He gained that perspective from practical experience. It cleared his vision. "The farmer must either take up power or go out of business," he judged. "Power farming is simply taking the burden from flesh and blood and putting it on steel."

In his 1926 book, *My Life and Work*, he reflected, "The automobile is designed to carry; the tractor is designed to pull. The public was more interested in being carried than in being pulled; the horseless carriage made a greater appeal to the imagination." So, with regret, he put aside development of tractors to work on automobiles for a while. But he waited only until the number of automobiles in the world sparked the demand for better, cheaper, lighter, and more manageable tractors.

For Ford, the future of the farm tractor pulled sharply away from the past. The large, heavy steam tractors grew out of the theory that weight meant power. Yet, if tractors were to pull rather than carry, excess weight detracted from the capacity to pull. Ford reasoned that his tractor had to be light yet strong, simple to operate, and yet cheap to buy and maintain. Cost was important. Anyone who wanted a tractor should be able to get one. Having and using it with ease was a key to his greater goal of making food and clothing more affordable as well.

It seems, in retrospect, that Ford thought and spoke in eloquent sound bites; it is close to true. He was one of North America's most public industrialists. Born in 1863, he was the second generation Ford in North America born onto farmland outside Dearborn, Michigan. Grandfather John emigrated from his own small farm near Cork, Ireland, in 1847, to settle near

▲*With 3.44x3.6in bore and stroke, the new NAA engine provided a peak 20.2 drawbar horsepower at 1,750rpm and a maximum of 30.2hp at 2,000rpm on the belt pulley. Even with a twenty percent increase in power, fuel economy remained 11.2 horsepower hours per gallon.*

▶*By the time Henry Ford celebrated his fiftieth year in business during 1953, peace and tranquillity had nearly returned to his life and his business. The break up with Ferguson was nearly resolved, and his Golden Jubilee was a new, improved machine. A 4in longer wheelbase, 5in longer overall, and 100lb heavier, it also boasted a new 134ci overhead valve engine.*

relatives in Michigan. He bought 80 acres for $350. Henry's father, William, added land and expected his son, Henry, to follow the family tradition. But to Henry, farming was drudgery. He preferred to tinker with machines. In 1896, his first machine ran, and by 1903, his motor company had begun. Ford started his first serious tractor experiments late in 1905. He established a satellite "tractor works" just three blocks from his main plant. He worked through several tractor attempts. Then in 1907, a prototype using the Model B engine was sent to his nearby Fair Lane farm for testing. Two more prototypes emerged and were tested through the fall and winter. In 1910, Ford applied for several patents even though tractor manufacture had taken a back seat to the automobile. His board of directors was unimpressed. The board disapproved of his plans to manufacture tractors at the new Highland Park plant. So in 1915, he left his own company over these disputes, and incorporated his new tractor company, Henry Ford & Son, in July 1917. (The Ford name was registered by Minneapolis businessman Baer Ewing, who produced a tractor, the Ford Model B, hoping to capitalize on the confusion. See chapter three.)

Ford collected samples of nearly every competitors' tractors. He tested them on his farm, then gathered them at the new Dearborn plant for his staff to examine. Because efficient manufacture was as important to Ford as product affordability, the tractor was designed in three units: a transmission housing contained the gearbox, differential, and the worm and worm wheel drive to the rear axle; the engine included flywheel and clutch assemblies; and the front end included mounts for the steering assembly and axle. Ford intended the tractor parts to be strong enough to support the entire ma-

▲*Harriet Fossum sets off to work while husband Palmer looks on. Much like what occurred at thousands of other farms, Harriet learned to operate a tractor when she was ten, taught by her father. She operated a grain binder and a bale loader. When Palmer went off to the Korean War, Harriet—like millions of soldier's wives before her—joined the work force. She worked nights as a nurse while, all around her, women kept farms operating and food on the tables throughout the U.S. and Canada.*

▶*It's simple if you know what it means. Drawbar horsepower peaked at 1,750rpm yet the engine could handle belt loads up to 2,000rpm. The needle indicated ground speeds in each gear at an indicated engine speed. Top speed in fourth gear was quoted by Ford to be about 11.5mph, at roughly 1,750rpm.*

chine without needing a separate frame. In his factory, all these units were designed to be run on rails to a central point for final assembly. He had used vanadium steel in the Model T, but chrome carbon steel was introduced for the tractors.

The first production run of fifty went back to Fair Lane farm for testing. Word leaked out. Stories appeared in magazines and newspapers. The world knew Ford's car; now it learned that his tractor was coming.

World War I had begun in Europe. The Germans were sinking a ship a day. England was losing its farmers and able-bodied draft animals and its food. With such losses, the British government sought help to encourage farm tractor production in the United Kingdom. Then on April 6, 1916, the United States entered the war. Ford sent Charles Sorensen, his assistant, to England with a carload of parts, patterns, and implements. He was to find a suitable factory, but at the end of June, the Germans bombed London's Fleet Street financial district. All the possible tractor plants were rushed into war plane manufacture; the tractors had to come from the United States.

Cables went to Ford. The British government ordered a minimum of 5,000 Fordsons at cost-plus-$50, about $700 each. First delivery was to be within sixty days! Sorensen returned home. The Dearborn plant was not yet ready, and the facility at Highland Park was working at capacity on war materiel. Later, limits in available shipping space slowed delivery. But somehow, Ford found a way. By early December, Fordsons began to arrive in England. A total of 7,000 were there by the spring of 1917.

Britain had lost 350,000 farm hands by the time the Fordsons arrived. The food situation was desperate, and the government

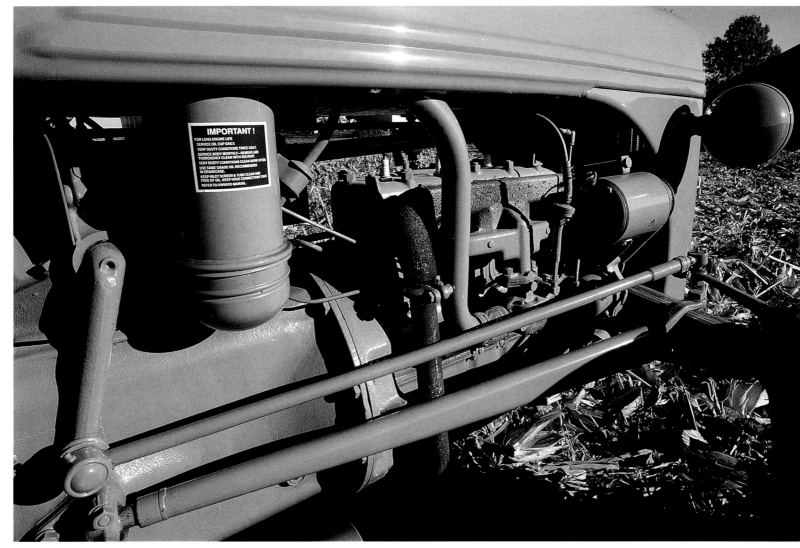

▲ *The nomenclature of the Ford-Ferguson tractors is fairly common knowledge: the 9, 2, and 8 refer to 1939, 1942, and 1948 as introduction years for the series. The 2N designation came as a corporate decision regarding governmental restrictions on raw materials at the beginning of World War II. Only by introducing a new model could certain supplies be made available or could prices be increased. (Deere & Co. did something similar with their Model G and GM.) This is the 1942 Model 2N.*

asked farmers to cultivate an additional 500,000 acres in 1917.

In Ireland, where most of this new land was to be opened up, Harry Ferguson, an aviator and auto racer, had already taken the tractor into his heart and his business. He had begun selling the Overtime, the London-built Waterloo Boy.

Ferguson, the fourth child of James Ferguson, was born in 1884, and raised on the family's 100-acre farm, 16mi south of Belfast. As with Ford, Ferguson did his farm chores, but anything mechanical easily lured him away from the farm. He was saved when his oldest brother, Joe, hired him to apprentice at his automobile workshop in Belfast.

Ferguson found himself. Machines became his passion. He even built his own airplane. He succeeded in every endeavor through salesmanship and showmanship. By the time the war in Europe approached the United Kingdom, he was in business for himself, and the need for food married Ferguson's work with his enthusiasm for machines. He jumped into tractor demonstrations with his Overtime. As a result of his successes, the Irish government approached him. To improve tractor performance nationwide, they asked him to visit farmers to perform educational shows. While doing these shows, Ferguson saw mostly tractors imported from the United States. While he found many of these cumbersome to use, what concerned him was the inefficiency and danger in the single-point plow hitch.

With Ireland's soil, the risk of hanging up the plow on hidden rocks was great. Horses simply stopped moving. Tractors with a spinning flywheel tended to keep going. Often that motion simply wound the tractor around the final drive gear and brought the nose up. If the operator didn't quickly put in the clutch, the tractor would flip over. It took quick reactions to release the clutch. Ferguson saw farmers with slower reactions missing arms or legs when their tractors flipped. He met the widows of those not so fortunate as that.

Even if the impact did not flip the tractor, the plow was usually damaged after impact. Farmers were advised to use wooden shear pins to attach the plows, but turning a field for the first time could exhaust a large supply of pins. Another drawback Ferguson witnessed was that while using implements designed for horse-drawn farming, tractor farmers often did not get satisfactory results. Plows had their own wheels, a vast improvement over earlier versions that the farmers not only had to steer but also carry! These wheels also served to set furrow depth. But over uneven soil, with the drawbar rising and falling as the tractor moved along, the furrow height either needed constant adjustment or simply ended up sloppy and uneven.

Harry Ferguson came to understand the tractor makers' motivation for great weight. It not only kept traction to the drive wheels, but it held down the plow. Nevertheless, he was con-

▸ *Harry Ferguson produced not only hitches but a full range of plows to work off them. That was the single most significant difference between him and handshake-partner Henry Ford. Ford never saw the need for or any advantage in producing specific Ford-built implements coupled to his tractors. Ford missed income possibilities and received the blame when others implements failed to perform well on his tractors. Here, a Ferguson quarter-turn 16in bottom plow waits for work to begin.*

vinced there was a better way.

Back in Dearborn, Henry Ford & Son tractor company was busy. Within three months of completing the British shipment, Ford had delivered more than 5,000 of the backlog of some 13,000 North American orders. By April 1918, daily production was sixty-four. In July, 131 Fordsons rolled out of the Dearborn plant every day.

It didn't take much time to uncover the Fordson's shortcomings. The early design set the worm and worm gear right below

▲ *Few tractor makers had as many aftermarket accessories available as Ford, again because of his unwillingness to make his own. This 2N, owned and restored by Dwight Emstrom of Galesburg, Illinois, has aftermarket steel sand lugs mounted outside the rear rubber tires as well as an aftermarket clutch mechanism that quickly disengages the clutch and operates both brakes if the plow hits a rock.*

▲The Ford 8N engine, tested in a tractor at the University of Nebraska in June 1950, measured 3.19inx3.75in. bore and stroke with a total capacity of 117.9ci. It was rated to run 1,750rpm at which it produced 17.7hp on the drawbar. Ford allowed engine speed to increase to 2,000rpm for belt pulley work and, in that application, the engine produced a maximum of 25.5hp with fuel efficiency measured at 11.2 horsepower hours per gallon consumed.

▼This was an early industrial application whose identity has long since rusted away. The 8N tractor engine was mounted onto channel iron rails and set on wheels. A clutch was fitted to the rear at the flywheel, and this was attached directly to the pump impeller. Performance and pump capacity cannot even be guessed, but it must be presumed that, if necessary, the engine moved a lot of water when it ran at 2,000rpm. This unique 1952 8N-6 Ford is owned by Palmer Fossum of Northfield, Minnesota.

the farmer's seat. These gears were inefficient, generating as much heat as power. This heat transferred up to the farmer's steel seat. Later versions inverted the system and bathed the worm in oil, cooling the system and the farmer's backside. By 1920, Fordson's problems had largely been solved and sales, always steady, reached 70,000 near year end.

From his earliest tests, Ford hoped for a tractor-with-plow as a single unit. Current thinking still reflected the historical: The plow was hitched to the horse. No one thought that the plow could be part of the tractor. This was a new concept, as unit construction had been.

In Belfast, Ferguson had been working on such a plow. Using a Ford Model T with the Eros tractor conversion (among the better of the dozens of conversion kits available, this one came from St. Paul, Minnesota), he had a lightweight "tractor." Ferguson's wheel-less plow was nearly ready. In December 1917, he learned that a Ford tractor plant was planned for Cork. Until production was running, several thousand more Fordsons were to be imported.

Ferguson the showman looked at this as access to a much larger market for his plow. He grabbed his drawings and raced to London to meet Sorensen. Ferguson biographer Colin Fraser set the scene in his 1972 book, *Harry Ferguson: Inventor and Pioneer*: "'Your Fordson's all right as far as it goes,' Ferguson told Sorensen, 'but it doesn't really solve any of the fundamental problems.' Ferguson got Sorensen's attention. He unrolled his drawings and explained his theory that achieving efficient farm mechanization lay in equipment designed on the unit principle—that is to say the implement becoming part of the tractor when it was hitched on, but being readily detachable again."

This was exactly what Ford had said. And Sorensen's recollection in his biography in 1956, *My Forty Years With Ford*, was more that he had suggested Ford's idea to Ferguson. The unit idea ultimately changed tractor farming for good. From the start, Ferguson had shared with Henry the ambition of easing the farmer's workload, based on the same unpleasant experiences from childhood. Ferguson persevered, his greatest skill after engineering and salesmanship. The result was the Duplex hitch, a design beautiful and simple in its engineering. It was completed by the end of 1917. Viewed from the side, if the lines of the upper and lower arms were extended, they would have met several feet in front of the nose of the tractor. The effects of physical laws on this triangulated Duplex hitch meant that the drag on the plow always created downward pressure on the tractor. The greater the drag, the greater the down force.

An additional benefit of Ferguson's system was that not only could a lightweight tractor be used, but also the plow itself no longer needed to be of great weight. Its only drawback was that, as the tractor pivoted over changes in the field surface, the depth of the furrow changed opposite to what the front wheels did. There was no draft control.

After Sorensen's first meeting with Ferguson, both agreed to keep each other informed of developments. In 1920, Ferguson arranged a demonstration in Dearborn of his plow system to fit the Fordson.

The demonstration was a success, up to a point. The Ferguson plow performed admirably, but Ford and Ferguson had different goals in mind, different assessments of each other, and different views of their importance to each other. Ford thought Ferguson was an innovative and effective machinery salesman. He told Sorensen to hire him on the spot. Ferguson had no interest in working for anyone else. He wanted Ford to back him in a plant in Ireland. Ford dismissed the whole thing but offered to buy the patents from Ferguson. Ferguson politely refused. Both Ford and Ferguson knew that without a depth-control device the plow's usefulness was in doubt. Afterward, still in Dearborn, Ferguson contacted Eber Sherman, a New Yorker who was Ford's distributor for South America. Sherman agreed to handle sales of the plow and to help Ferguson find a manufacturer. Ferguson sailed home.

Ferguson's lack of success with Ford hurt. It brought his backers to their feet and knocked the inventor to his knees. After bitter disagreement, Ferguson resigned from Harry Ferguson, Ltd., and opened a new shop elsewhere in Belfast.

Back in the United States, Ford's Fordson was a continuing success.

Ford had entered the tractor business in 1917, and by 1921, he had hold of two-thirds of the entire market.

Ford had also succeeded in reacquiring majority interest in his automobile company, and he moved Ford & Son tractors into his new River Rouge plant. His Dearborn plant production had reached a record 399 a day, 10,248 for September. But the war ended. The River Rouge plant worked at capacity. It was too much. Overcapacity caught up. Sales in the 1920-21 depression dropped to 36,793 (though this was partly caused by new plant set-up time).

To keep production up, Ford cut the price of his tractor. Then he cut it again and again as he searched for the price level that would put the tractor in everyone's hands and keep his factories busy. Production had risen to nearly 69,000 in 1922, despite the economy, and to nearly 102,000 in 1923. A healthy export business continued, and between 1920 and 1926, nearly 25,000 were delivered to Russia. He took losses to meet those goals. The price war that resulted enraged and broke many competitors. But he had not only temporarily overwhelmed his competition, he had educated them.

In her book on International Harvester, *A Corporate Tragedy*, Barbara Marsh assessed Ford's impact. "The Fordson, much lighter than other tractors on the market, exemplified Detroit's know-how. It proved Harvester's engineers had a lot to learn about the refinements of internal combustion, heat treatment of steel, strength of materials and standardization of parts. Ford's example behooved Harvester and old-line makers of farm equipment to revamp their ancient production methods for the precision-machining requirements of the tractor. Reaping economies in production, Ford shattered the tractor industry, already undergoing a shake out, with repeated price slashing. By 1922, when the price on a Fordson dropped $230 to a level of $395, Ford effectively broadcast his willingness to lose money on tractors to keep production going."

International Harvester introduced its Farmall Regular as a direct result of Ford's Fordson. War was declared in the farm fields, and IHC's engineers went right to the front lines to test and improve their product. Implements dedicated specifically to the Farmall were introduced, and IHC's full-line dealers sold the system. Ford, still selling largely through his automobile dealers, had never recognized the value of his own implements.

Not only was there a profit incentive, but implements specifically meant for Fordsons would have better shown off the tractor's capabilities. Instead, farmers around the country made do

◄The ARPS tracks were all-steel offered in 13in (recommended for plowing to fit within the 14in furrow) or 16in widths. The track shoes were pressed steel with 1in grousers. A pneumatic rubber idler wheel kept track tension, using a 5.0x15in four-ply implement tire. Track width was adjustable at 52, 56, or 60in centers. Between 1920 and 1940, Arps produced more than 14,000 crawler and half-track kits.

▲Since he retired from active dairy farming in the early nineties, Palmer Fossum of Northfield, Minnesota, has made it his new job to gather and collect the most unusual pieces of Ford tractor history. His 1956 Model 650 is fitted with a Ford adjustable plow and the Blackhawk Half-Track manufactured by ARPS Corporation of New Holstein, Wisconsin.

▲The Tractor and Implement Division of Ford Motor Co., of Birmingham, Michigan, introduced the 600 series of tractors in 1955, to replace the two-year edition of the Jubilee. At the same time Ford also introduced the larger engine displacement 800 series. The 650 used Ford's 134ci inline four carried over from the Jubilee and hence, not specifically tested at Nebraska. However, the new tractor offered a five-speed transmission.

with Cockshutt, Oliver, or Deere plows or other attachments. When the Fordson failed to perform as advertised or expected, the mix of manufacturers was not blamed, the tractor was. Despite ever-falling prices, Fordson production declined, and by early 1928, Ford quit selling Fordsons in the United States. International Harvester had the lead again.

When Lord Percival Perry, head of the British Ford Motor Company, stopped in Dearborn on his annual visit, he saw stockpiles and silenced assembly lines. He seized the moment. He proposed to Sorensen (who was by now the head of tractor operations at Ford) that he, Perry, take the machinery and dies back to the United Kingdom. Perry would use them there to build tractors where they were still needed. Ford, who wanted development to continue on the unit plow and who preferred the automobile line had shut down instead of the tractor plant, agreed.

Ferguson continued his developments and even visited Ford several times during the twenties. He had nearly solved the draft control dilemma. A "floating skid" patent applied for in December 1923 worked in the interim. But still not satisfied, Ferguson adopted an internal on-board hydraulic-lift system to ease the chore of turning the tractor at the ends of rows. By adapting a sensor to the hydraulics, by replacing more rigid mounts with

▲*Palmer Fossum mows his farm yard outside Northfield, Minnesota, with his Powermaster and a PTO-driven John Deere three-blade mower. The 1957 801 measured 66in tall, 72in wide, and 133in overall on an 85in wheelbase. It operated on 5.50x16in front tires and 12.0x28in rear tires.*

ball joints, and by increasing the angle of the top strut (now a single instead of a pair), he closed in on the perfection he sought.

In 1925, he had incorporated in the United States with Eber and George Sherman, as Ferguson-Sherman, Inc., to manufacture and market his plow. In 1928, his new three-point hitch with automatic draft control was ready. Yet Ford was switching over his Dearborn plant to production of his new Model A automobile. Ferguson was caught again without a manufacturer. And then it was 1929. The stock market crashed, and the Great Depression began.

Ferguson did not give up. Reasoning that, in tight money times, it might be easier to sell something if a manufacturer could see it in the flesh (or at least in steel), Ferguson ordered parts to build a prototype tractor and unit plow. Transmission pieces came from Britain's largest gear producer, David Brown Co. The engine, a four-cylinder Hercules, came from the United States. The tractor began to come together in Belfast, and when complete, Ferguson ordered it painted black. Some say it was because of Henry Ford's dictum that only black was suitable for cars. But it is equally likely that Ferguson, who had so little tolerance for unnecessary frills, looked on black as the most proper, unadorned finish. It was only later, seeing how field dirt contrasted against the black, that he changed to gray paint.

The black tractor went into tests, but a draft control problem leftover from the Fordson still existed. An experiment proved that instead of using the hydraulic cylinder oil under pressure on the bottom two links to position the plow and control its draft, it worked better to use the hydraulics on top to lift the plow by compressing the cylinder and letting gravity take the plow down.

Ferguson had succeeded. Now a manufacturer was needed. Ferguson took to the road again, this time staying in the United Kingdom. Gear maker David Brown became a tractor manufacturer, and the first Brown-Ferguson machines rolled out in early 1936. Demonstrations quickly quieted farmers' skepticism in the United Kingdom.

Ferguson invited Eber Sherman to one of his demonstrations. He hoped that Sherman would tell Henry Ford about the new machine and Ferguson's hopes for vast worldwide production.

"What the world needs right now," historian Allan Nevins quoted Henry Ford in the fall of 1937, "is a good tractor that will sell for around $250." Sherman returned to Dearborn as Ferguson's emissary for just such a machine, and in late 1938, Ferguson and a small staff took a Ferguson-Brown to Fair Lane. Ford had only reluctantly ceased tractor production ten years earlier. Fordson production continued first in Cork, then after 1933, in Dagenham, England. Yet Ford was thinking of a new model for

North American sales. Several models had been developed, one a three-wheel row-crop type with Ford's flathead V-8; another was a four-wheel standard in which Ford wanted an overdrive road gear. But Ford was dissatisfied with his own engineers' work.

Ferguson's timing was perfect. His new tractor finished its demonstration. Ford was quiet, but soon offered again to buy Ferguson's patents. Ferguson replied that Ford didn't have enough money and they weren't for sale anyway.

It remains one of the great business stories of modern history that these two men, each stubborn and distrusting of written contracts, would consummate a deal so grand and important out of doors, in private, out of earshot of witnesses, and simply on a shake of the hands.

A gentlemen's agreement. Colin Fraser reported the terms: "Ferguson would be responsible for all design and engineering matters; Ford would manufacture the tractor and assume all risks involved in manufacture; Ferguson would distribute the tractors, which Ford would deliver; either party could terminate the arrangement at any time without obligation to the other, for any reason whatever, even if it was only 'because he didn't like the color of his hair'; and the Ford tractor plant in Britain [at Dagenham] would ultimately build the Ferguson System tractor on similar terms to those established for Dearborn."

All agreed to and sealed with a handshake.

Ferguson was overjoyed. Ford was excited. But it was a short honeymoon. Sorensen's recollections were clear. "When Ferguson appeared in 1938, we were ready with a new tractor, and his plow with a hydraulic lifting device appealed to Mr. Ford…We wanted him to adapt a plow to our tractor. We did not need him to show us how to build tractors—he needed us. We did want him to come with us because we knew we would make a success of his plow if we could adapt it."

Allan Nevins' history emphasized it was Ford, not Ferguson, making design decisions and specifications. Ford invested $12 million in tooling costs and helped Ferguson finance his new distribution company, Sherman having dissolved Ferguson-Sherman Manufacturing.

The 9N, the Ford Tractor with Ferguson System was introduced June 29, 1939. Its $585 price included rubber tires, power take-off, Ferguson hydraulics, electric starter, generator, and battery (lights were optional). The gasoline four-cylinder had an automobile-type muffler, and 9N sales brochures showed possible mounting points for a radio, due to the quietness of the engine.

Ford's 9N improved upon the cantankerous Fordson by updating the ignition with a distributor and coil. An innovative system of tire mounts for the rear wheels and versatile axle mounts for the fronts enabled farmers to accommodate any width row-crop work they needed, from 48 to 76in, using nothing more than the supplied wrench and jack. This may have been a Ferguson invention since it disappeared from the 8N tractors. But then, so did the Ferguson System badge on the Ford's radiator.

Ford aimed for the perfect tractor with his 9N. He had tried before with the Fordson and believed he had it right by the time he introduced his new model in 1939. Still, there were problems with his newest effort, some of them matters of operator comfort and convenience. With the introduction late in 1942 of the 2N, several changes occurred because of wartime needs for metal and rubber. Chrome instrument bezels, battery covers, and radiator caps all went to steel in '41.

Ferguson again was a victim of timing and a new world war caused orders for the 9N to reach little more than half the original projection: 35,742 instead of 63,750. Production nearly touched 43,000 in 1941, but wartime rationing of rubber returned all tractors to steel wheels. The lower output raised Ford's costs, and eventually the price followed.

Still, until the 2N arrived to replace it, the 9N made significant impact on the farmer, as Ford and Ferguson had hoped. And it demonstrated the engineering ideas Harry Ferguson had struggled to prove for decades.

Just before the Second World War, more than 6.8 million families in the United States still lived on farms. Only 1.2 million had tractors while, according to a U.S. Department of Agriculture study, 17 million horses still worked the farms. Nearly one-fifth of the land under cultivation in North America was devoted to feeding the draft animals, which needed to eat year round, a fact recited regularly by Ferguson and Ford.

In 1948, Ford introduced the 8N. Ferguson's name was nowhere to be found. Known throughout England as a feisty, opinionated inventor, his outspoken nature and his strong belief in his correctness put many people off, some of them in positions to do him great good. He had rankled Sorensen. The directors of Ford Motor Company, Ltd., refused to seat him or to manufacture his tractor while their own Fordson continued to sell well.

The "Handshake Agreement" proved as good as the paper it was written on. Both sides eventually cheated on it.

When Henry's son, Edsel, died in 1943, the senior Ford, then eighty, returned to the company to run it. His grandson, Henry II, who was twenty-six, was called home from the Navy to take over. He learned that the tractor operation had already lost $20 million. (The U.S. Office of Price Administration had allowed no price increases by any manufacturer during the war.) To watch expenses more closely, Ford returned tractor operations to Highland Park from the Rouge plant.

At the same time, Henry Ford hoped to clarify his agreement with Ferguson. After frustrating Ford the man and Ford the company, Ferguson returned home and left things for his company's president to handle. Sorensen had characterized Ferguson as a "stand-by consultant" to Lord Perry. In 1946, Ford Motor Company tried to buy into Ferguson or buy him out, to form a new sales and distribution company. Ferguson's temper rose at the final offer, thirty percent of the new company and no royalties on his patents.

Negotiations broke. Effective December 31, 1946, Ford's agreement with Ferguson would end. Ford was to continue to manufacture tractors through June 1947 for Ferguson to market, but Ford immediately established its own distribution company, the Dearborn Motor Corporation.

Then Henry Ford died on Monday, April 7, 1947, at age eighty-three. In his lifetime, he had sent 1.7 million tractors out of his factory doors with his name on them.

The refusal of Ford Motor Company, Ltd., to manufacture Ferguson's tractor had led him outside by this time. A 2N clone, the Ferguson TE-20, went into production in Coventry, England. It used a Lucas electrical system, and its engine had overhead valves and a four-speed transmission, improvements Ferguson advocated for the 9N and 2N but which Dearborn had not yet picked up.

In July 1947, right after Ford's last shipments to Ferguson, the new 8N was introduced. It boasted some twenty improvements over the 2N, including a four-speed transmission. It came equipped with the Ferguson System. No royalties were paid.

◂The 801 Powermaster-series tractors were three- or four-plow rated machines. These were offered as gasoline, LPG, or diesel-engine versions. A twenty-amp generator recharged the six-volt battery and electrical system. Electric lights were standard, and an optional lighting kit put implement lights on the rear fenders.

◂The "Red Tiger" Powermaster engine had a bore and stroke of 3.90x3.60in for total capacity of 172ci. Ford used the same block for LPG and diesel fuel as well. Engine output was 31.9 drawbar horsepower and a maximum of 45.7 brake horsepower. A five-speed transmission was used that provided a top speed in road gear of nearly 12mph. Power steering was standard.

▸▸This 600-series Ford was equipped with only a single-point hitch. Of course, this one also had a transmission oil cooler, front and rear fenders, a full cab with white reflector tape, and a rotating amber safety beam on the roof. This was not standard equipment in reaction to the Ford-Ferguson breakup. It was meant for use by the U.S. Air Force an airplane tug. It is difficult to know now just how many were produced but the "Moto-tug" was common throughout U.S. Air Force and National Guard bases through the sixties.

◂ *This three-door cab—one sliding door on each side plus a hinged door at the rear—provided seating for three as well. Two tool boxes beneath the rear windows served as jump seats. A torque-converter transmission made best use of the maximum 22.4 drawbar horsepower from the Ford Industrial four-cylinder 134ci engine. A huge heater/defroster kept the operator warm and the windows clear through the winter. This curious Ford is another part of the collection belonging to Palmer Fossum in Northfield, Minnesota.*

Ferguson immediately set out to produce a tractor for the United States. His TE-20 was to have a cousin in the colonies, the TO-20. But his financing dried up. Potential investors worried over Ferguson's ability to compete against the power and size of Ford. He approached other car makers, Willys-Overland, Kaiser-Frazer, and even General Motors.

But GM disagreed over the size of the tractor the U.S. farmer needed, and Willys would agree only if it had controlling interest. Ferguson, more and more riled by this time, directed his wrath at Ford.

"It'll be a grand fight" Harry Ferguson said, as he filed suit against Ford. He claimed damages of $251 million. He charged "conspiracy" to infringe on patents, to willfully destroy his distribution business, and to block manufacture of its own tractor. Ford denied or repudiated every allegation. The legal battle, now not merely a patent suit but also an antitrust suit, dragged on for four years. More than one million pages of evidence were taken, nearly 11,000 from Ferguson himself.

Meanwhile, Ferguson began producing his TO-20 in Detroit. By the end of 1948, he had recorded a profit of more than $500,000; 100 tractors a day rolled out the doors. So Ford counter-sued Ferguson in July 1949, charging him with "conspiracy" to dominate the world tractor market and ruin Ford Motor Company. Ford hoped to wear Ferguson down. The opposite occurred. When the suit came to trial in March 1951, Ferguson added $90 million more to cover the Ford tractors built since filing the suit.

The trial dragged on for months. It did wear Ferguson down. On July 17, 1951, he told his lawyers to accept a settlement if Ford offered. Ferguson, exhausted, had spent $3.5 million and received $9.25 million on April 9, 1952. Ferguson's patents on much of the three-point hitch had run out by the time the suit was settled. The remaining pieces still covered had to be redesigned by Ford. But Ford was already at work developing its next tractor.

Ford's fiftieth anniversary in 1953 was commemorated with the new NAA tractor that began production shortly after the new year began. Substantially restyled, it was officially named the Golden Jubilee 1903-1953 Model but quickly became known as the Jubilee. The 8N had grown some to become the Jubilee. It was 4in longer and higher and about 100lb heavier. It also introduced Ford's new Red Tiger engine, an overhead-valve four. A new vane-type hydraulic pump replaced the Ferguson System pump, and it was relocated to the right rear of the engine. Live power take-off was optional. The Jubilee, a three-plow tractor, was produced until 1955, when Ford changed its tractor direction.

Until the introduction of the new 600 and 800 series, five tractors in all in 1955, Ford had been a one-tractor company since 1917. Now Ford, which had on and off again held the major market share, could compete more effectively against all its opponents.

LPG, liquefied petroleum gas, became a fuel option in the United States and United Kingdom, and diesel engines were available for the Fordson Major. Workmaster and Powermaster tractors were introduced in 1958, with model series numbers from 601 through 901. Diesel engines also came to the United States in 1958. In 1959, Ford introduced its "Select-o-Speed" transmission on its 881 model, using hydraulic power for gear change in an automatic-type transmission.

In the year of Ford's Golden Jubilee, Harry Ferguson fulfilled his childhood wish of immigrating to Canada. In 1953, he sold his tractor and plow company to Massey-Harris in Ontario. Automobiles once again piqued his interests, and most of the proceeds of the sale, $16 million, were invested into acquiring rights for the torque converter and all-wheel-drive systems.

Harry Ferguson died on October 26, 1960. Less than three months before, he had contacted friends about a new idea—about getting back into the game. He wanted to build a new tractor, a tractor that would make use of the torque converter automatic transmission and four-wheel drive.

Chapter Nine

A Brief History of Tractor Innovation

Design and Engineering Broaden Their Impact

It took the human race thousands of years to advance agriculture beyond animal power. Beginning in the mid-nineteenth century, the inventors and tinkerers of the previous 200 years would have seen their ideas put to practical use. But the next evolutionary cycle took only a fraction as long. It was not even seventy-five years before things changed again.

From the late 1880s to the late twenties, the farm tractor manufacturing industry evolved steadily. It switched from primarily steam engine power for stationary, portable, and traction engine applications, to the gasoline engine. The importance of engineering became clear to manufacturers as they changed their products from steam to gas. With the arrival of tractor testing, the role of the engineer was enlarged even more. The buying public began to recognize its own power. Manufacturers could, of course, provide a "ringer" tractor, a hopped-up, cheater machine for the tests. But results achieved on the dirt test oval at the University of Nebraska were expected in the potato fields of Maine and wheat lands of California. Predictability and dependability were performance characteristics that advanced tractors beyond the capabilities of backyard tinkerers and early-day blacksmiths.

The next change was even more dramatic, though the machines looked little different. Eleven years after the first Nebraska tests, Caterpillar demonstrated reliable, practical diesel power. Soon after its 1931 introduction, just as with gas and steam before it, diesel power became available for stationary, portable, and tractor applications. However, the engineering of the diesel was even more critical. The difficulty of starting engines using the gelatinous fuel on a cold morning necessitated pure engineering solutions. It took some compa-

▲*The YT used one-half of a Minneapolis-Moline four-cylinder engine to make a two-cylinder engine. Crudely cut cylinder heads topped the three prototypes while this head is more finished in appearance with head bolts being further into the casting than at the edges in the prototypes. The carburetor was a Schebler TR while the magneto was a Fairbanks Morse Model FMJ.*

▶*The YT measured 80in tall to the top of the stack, 80in wide, and 119in long overall on a 79in wheelbase. The transmission was a four-speed-forward/one-speed-reverse gearbox similar to later Model Rs. Front tires were 7.50x10 while rears were 9.50x38s. Engine specifications are pretty much unknown as it appears that nearly all but a few of the twenty-five versions built were recalled to the factory and probably destroyed or adapted to other experimental units.*

nies until the late fifties to feel confident enough of their engineering and trusting enough of farmer demand for it to offer diesel engines. By this time, some of the diesel pioneers had already phased out all the other fuels from their product line-up.

After diesel engines, the next step in the evolution was much more noticeable; yet, it occurred within seven years of the new power. A new breed of engineer became involved—the industrial designer. These people were part sheet metal designer, part manufacturing specialist, part marketing researcher, part aesthetician, part structural engineer, part agricultural engineer, part orthopedic surgeon, part magician, and part miracle worker. At first, this amalgam was brought in simply to make one manufacturer's tractors stand out from all the rest, to become "more salable" than the competition.

In 1938, Deere & Co. introduced "styled" tractors. By 1950, every major company manufacturing tractors in North America had hired industrial designers and engineers as outside consultants or as inside staff.

In *Webster's New World Dictionary*, an "engineer" is described, among other ways, as "a person who operates or supervises the operation of engines or technical equipment; a specialist in planning and directing operations in some technical field." That these individuals began to work on tractors is obvious. It is the definition of "engineering" that is more apropos: it is "the science concerned with putting scientific knowledge to practical uses…; the planning, designing, construction, or management of machinery…, etc."

This is most accurately what happened to farm tractors from the early 1900s through the early sixties.

Alvin Lombard's Log Hauler, his turn-of-the-century steam traction engine crawler, fed its steam from the boiler equally to each side's compound, two-cylinder,

◀ *The 1937 YT was a curiosity. Three prototypes were built as development mules before a limited production run of another twenty-two was completed. Each was given a U.S. Army Air Force serial number, this one being #52815. But whether it was meant to be an aircraft tug or something else, it never went into regular production. This experimental tractor was restored and is owned by Walter and Bruce Keller of Kaukauna, Wisconsin.*

◀◀ *Experimental models are prized by collectors for their rarity and mystery. Each company builds models it never puts into regular series production. For reasons of inadequate performance, insufficient interest or demand from the marketplace, or just a simple change of mind, the limited edition tractors are tested, prodded, measured, and, then, all the evidence of their existence is generally destroyed.*

vertical engines. There was only one throttle valve for all four cylinders, located in the dome. There were no track clutches or track brakes to slow one side to assist in turning the 19ft-long, 18-ton, 100hp engine. It was all accomplished by the tillerman on the wood bench mounted above the front skis. Freezing in the winter, sitting ahead of the boiler, he cranked the worm-and-sector steering gear to change direction of the lumber train behind him. Lombard's Log Hauler had no compressor or air brakes. He expected that friction—even on ice—would stop the crawler and its loads even when tugging sleds loaded with twenty tons of logs each. When Ben Holt adopted the Lombard crawler technology, he added track clutches and brakes to more efficiently maneuver, or to stop, his 23ft-long, tiller-wheel, gas-engined crawlers.

The English tractor, the Field Marshal Diesel, was started simply. One method used a shotgun shell slipped into the cylinder head and triggered when the cylinder was full of fuel and just past top dead center. But this was hard on main bearings. The other technique called for removing the T-handle near the head (it's hollow on the inside end) and placing a wick soaked in diesel fuel inside it, setting a match to the wick and letting it burn until it glowed hot. Inserting it back into the engine, the operator opened the compression relief valves and started pulling on the massive 24x5in flywheel. This method needed two strong men. Once the compression built up, the machine would pop to life.

Deere's diesel, and those that came before it and since—Caterpillar, IHC, Allis-Chalmers, White, and Ford—accomplished the same complicated tasks much more easily. Deere and Caterpillar mounted a small gasoline-powered engine on the diesel engine block. Its exhaust was vented through the larger diesel block to preheat the cylinders, and then, when the larger engine was sufficiently warmed, the smaller gas engine was engaged by clutch to turn over the diesel, building compression until it popped to life.

Power take-off (PTO) was at first reliant on tractor forward motion for its power shaft drive to rear-mounted implements. When it first appeared in 1929 as independent, or "live," PTO on International Harvester Model 10-20 prototypes, it was a complicated, cumbersome, dangerous, and ugly affair, operated by a clutch off the front of the crankcase. A pulley wheel ran to a chain that drove a shaft along the side of the engine. The shaft ran back to a bevel gear at the belt pulley that was connected internally to the PTO shaft. With the front clutch engaged, the belt pulley and the PTO shaft ran at the same time. IHC never marketed it.

It would be another eighteen years before live PTO would be commercially available. Engineered and introduced by the Canadian manufacturer Cockshutt for its 1947 models, their system coupled the PTO shaft to its own clutch within the transmission housing. No longer would harvesters or binders or any other shaft-driven implement slow or stop its function when the tractor slowed or stopped.

World War II interrupted regular production as tractor makers were drafted to produce tanks and artillery shells and guns. Just as during World War I, a variety of metals, particularly bronze, copper, chromium, and steel, came into short supply for civilian purposes. Tractors and implements that might have otherwise been discarded and replaced were repaired.

Recognizing that many rural males—more hired men than farmers—had gone to war, companies such as International Harvester organized programs and classes where rural women were taught to operate and maintain the equipment. Posters and advertising encouraged women to become the "field artillery," to become a "tractorette," as their urban counterparts had become Rosie the Riveter. Women hated the nicknames but rose to the challenge. One study in Kansas in 1943 showed women performing 85 percent of the farm machinery operation. Frequently farm women, adept at operating tractors, planters, and harvesters, hired townswomen to take care of their home, cleaning and preparing meals while they performed the field work for their absent husbands.

When the war ended, it took about a year to reconvert factories to tractor production. A great deal of equipment was back-ordered, and the establishment of the Marshall Plan only made matters worse. Existing, quality manufacturers worked frantically engineering ways to produce machinery faster. A number of new manufacturers jumped quickly into the business, shipping tractors overseas. Companies like General Tractor in Seattle, Intercontinental Manufacturing near Dallas, Implement Manufacturing Co. of Ogden, Utah, and Jumbo Steel Products of suburban Los Angeles appeared. Businesses were born named after founders, cities, days of the week, and emotional states: Long Manufacturing, Detroit Tractor with an all-wheel-drive model, the Friday Tractor Co., and Love Tractor, Inc., with a variable-speed machine capable of 42mph.

Companies popped up with odd configurations, unusual innovations, and unnatural engine choices. R. H. Sheppard brought out tractors with two- and three-cylinder diesels. Metal Parts Corporation produced its Haas "Atomic" with a one-cylinder airplane engine. National Implement Company offered its Harris Power Horse four-wheel drive. General Tractor Corporation manufactured its Powerbilt four-wheel drive that actually used the load shifting on the drawbar to help turn the tractor. Still others bore names that hinted at competence despite where they might be headquartered. Custom Manufacturing Co., National Implement Co., and the Earthmaster Farm Equipment Co. entered the postwar market boom, joining Farmaster Corporation with its offices in New York City and Global Trading Corporation with its offices in Washington, D.C. Each of these appeared with tractors that were of lesser or greater quality. But many of these disappeared soon after the major manufacturers caught up to back orders and began shipping overseas again.

Among major North American tractor makers, design and engineering improved the seats and reinvented the transmissions with developments like the torque amplifier on International Harvester tractors. Disc brakes appeared on Olivers, improved implement hitches and power take-off were offered on Cases with the Eagle Hitch, Allis-Chalmers introduced its Traction Booster, and four-wheel drive was finally tamed by the Wagner Brothers of Portland through the adoption of a key hinge.

▲*The Full Crawler Company, a department of George M. Smith Steel Casting Company of Milwaukee, Wisconsin produced only a handful of these conversion kits for Fordsons. A number of companies produced add-on half-track adaptations for the Fordson and a company called Snow Motors Inc., of Detroit, Michigan, even produced a pair of spiral cylinders as a replacement for wheels for use in deep soft snow.*

Aftermarket specialty makers introduced LPG conversion kits, do-it-yourself power-steering hydraulic systems for most tractors, and even turbocharger kits for many diesel makes. By the end of the 1950s, Harry Ferguson's three-point hitch had been adopted by the American Society of Agricultural Engineers and the Society of Automotive Engineers as the Category I standard system. This vindicated for all time any question of Ferguson's contribution to mechanized agriculture.

The end of the fifties was most notable for the first of engineering experiments with alternative engines. Ford startled the competitions' engineers and farmers alike with its experimental Typhoon, a free piston turbine engine coupled to Ford's Power-Shift semi-automatic transmission. This was first shown in 1957. Then in 1959, Allis-Chalmers showed off its fuel-cell tractor, an experimental model powered by what was basically 1,008 electric fuel cells, similar to wet-cell batteries.

Researchers at Iowa State University worked with Ford to produce a "tandem-hitch" tractor, which in its most manageable version mated one tractor without front axle or wheels to the rear of another in front of it. This system, first shown in 1960, provided more drawbar power, but conversion and operation was cumbersome. Deere & Co.'s solution to farmer requests for more power was an eight-year-long secret effort by engineers and industrial designers working out of a former grocery. In the end, Deere retired its long-cherished two-cylinder tractors and introduced four- and six-cylinder in-line, upright diesels in new sheet metal. At the end of an event- and technology-filled 1960, heavy-truck maker White Motor Co. of Cleveland, Ohio, acquired Oliver, overnight entering the tractor and implement business.

A year later, Allis-Chalmers introduced its first turbocharged diesel, its engineers controlling the power and the problems experienced by dozens of do-it-yourself engineers on the family farm. And International Harvester showed its experimental gas turbine engine coupled to its hydrostatic transmission. The startling-looking tractor, roughly the size of a Farmall 200, its body of fiberglass, was powered by an engine that weighed only 60lb, turned at 57,000rpm, measured 21in long, and produced 80hp. In early 1962, White added Cockshutt Farm Machinery to its holdings, and in 1963, it acquired Minneapolis-Moline. The same year, engineer Vernon Roosa, working in New York City, introduced his rotary-type, diesel fuel-injection pump to replace the much larger and mechanically more complicated in-line

▲ Ford's own in-line four-cylinder engine powered the Full Crawler version. Bore and stroke of the Ford measured 4.0x5.0in. At the rated speed of 1,000rpm, peak drawbar horsepower was 12.3 while belt pulley output was a maximum of 22.3. However, horsepower loss through the gearing and track drive would have been considerable. This 1924 machine is owned by Mrs. Edith Heidrick of Woodland, California.

209

▲The original jeep of 1938 resembled a U-DLX without the cab. Four-wheel drive was added by 1940, and the operator was moved more to the center of the vehicle. Minneapolis-Moline's Jeep measured 58in tall to the top of the steering wheel, 77in wide, and 166in long from front hook to the rear one. It was built on a 100in wheelbase. It rode on 9.0x10 10-ply truck tires. The four-cylinder engine used Delco electrics and a Schebler TR carburetor. This 1944 Jeep is owned by Walter and Bruce Keller of Kaukauna, Wisconsin.

▶While nearly every company produced Jeeps during the War, history seems pretty well set on Minneapolis-Moline producing the first vehicle called a "Jeep". While it was being tested at Minnesota National Guard Camp Ripley, the ability of this machine to go anywhere reminded some observers of a Popeye the Sailor cartoon creature with similar characteristics, called Jeep. The name stuck even when Ford built prototypes and Willys-Overland-built hundreds of thousands for the armed forces.

▲*Raymond Johnson of Fremont, Nebraska was famous for racing a Jumbo Simpson at county and state fairs! With a ten-speed-plus-overdrive transmission, he routinely averaged 60mph to win. He even was stopped once for speeding on a Nebraska highway at 70mph. The Jumbo used hydraulic brakes, necessary for Johnson's kind of driving. The 217ci Chrysler engine had bore and stroke of 4.25x4.375. At 1,200rpm, nearly 35hp was available on the drawbar.*

▶*With its front-hinged engine cover, servicing virtually anything up front was very easy. The Simpson ran on gasoline but could be fitted to operate on kerosene for $125. The PTO and belt-pulley were another $175, and a live onboard hydraulic system was $395. Jumbo also manufactured land scrapers and land levelers and other pre-planting implements. This tractor was restored and is owned by Raynard Schmidt of Vail, Iowa.*

▲*This tractor was manufactured in suburban Los Angeles by the Jumbo Steel Products Company of Azusa, California. Jumbo used a Chrysler industrial type 5A-210 engine that gave farmers throughout North America the benefit of going to a local Chrysler dealer for service. Jumbo's Model B row-crop tractor and Model C standard tread tractor sold for $2,050 in 1948. Jumbo produced its first tractor in 1946 and was out of the tractor business by mid-1955. This was the 1946 Model B.*

pumps that fed each cylinder individually with their own fuel pump. Soon after, Roosa introduced his "Master Pencil Nozzle," a smaller, simpler nozzle, better suited to the more prevalent direct-injection-type diesel engines that were overtaking the precombustion chamber head versions.

Following decades of complaints by wives and widows about farm tractor safety, Deere & Co. introduced the Roll-Gard rollover protection system, developed by its own engineers working intensely with Henry Dreyfuss & Associates. Unable to give it away to farmers who were too sure that tragedy would happen only to the other guy, Deere gave the technology to its competitors. No sooner was that safety issue addressed than concern over slow-moving vehicles surfaced. The Automotive Safety Foundation, monitoring the numerous rear-end accidents in which fast-moving automobiles hit slow-moving tractors, funded the creation and subsequent standardization of the fluorescent yellow and orange triangle, developed at Ohio State University by agricultural engineering graduate student Kenneth Harkness.

During this period, up through 1965—and the decades since then—Webster's definition of engineering has been a synonym for tractor development. Yet it has been the goal of everyone from Otto von Guericke to Nicolaus Otto to Rudolf Diesel to practice engineering: "The science concerned with putting scientific knowledge to practical uses."

▲*The row crop was delivered on 5.50x16in front tires and 10x38in rears. It stood 65in tall, 78in wide, and 129in long overall, on an 85in wheelbase. It weighed 3,000lb dry. The Dodge truck transmission offered five forward speeds with a top speed of 22.4mph. Rear track was adjustable. Jumbo used Chrysler automobile wheels and spindles on the front and a Dodge two-ton truck rear end.*

▲The 1948 Model E used a Wisconsin Model TF in-line air-cooled two-cylinder engine. With bore and stroke of 3.25x3.25in, total capacity was 53.9ci and Gibson rated the engine as 12.6hp at 2,000rpm. A Bendix 0.75in carburetor and a Fairbanks-Morse FMJ magneto were standard equipment. This little machine is owned by Paul Brecheisen of Helena, Ohio.

▶It's as small as it looks. It stands 72in tall to the top of the exhaust, 72in wide, 96in long overall, and its wheelbase is barely 68in. It weighed less than 1,900lb. It operated on 4.0x12in front tires and 8.0x24in rears. A three-speed transmission allowed a top speed of 7mph. Gibson Corporation was located in Longmont, Colorado.

▲Each set of front and rear stilts was virtually custom made although only two standard heights were available, 4.5 and 6.0 ft. The row width could be specified although it was not adjustable. Heat-treated 2.5in diameter stub axles were welded on to the rear stilts, and Timken bearings were used in steel hubs. Standard wheels were used. This 1948 Model B is owned by Walter and Bruce Keller of Kaukauna, Wisconsin.

▶In order to spray, dust, or defoliate cotton, or to cultivate, fertilize, or to top corn, most manufacturers left the farmer too low even with their high-clearance or hi-crop models. But Tractor Stilts Company, Inc. of Omaha, Nebraska, had an answer. The company produced its first stilts in 1948 and manufactured them for any tractor make or model.

◄ Live power-take-off was one of Cockshutt's claims to fame. This left the PTO shaft spinning independently of ground speed, enabling towed implements to continue their work even as the tractor might slow for the end of the row or stop altogether. Live—or independent—PTO did not originate with Cockshutt. It was first offered on Hart-Parr Model 18-36 tractors as early as 1926. PTO was first available on International Harvester Model 8-16 tractors introduced in 1918.

▲ Gambles—Cockshutt—used an in-line vertical four-cylinder Buda gasoline engine. The engine measured 3.43x4.125in bore and stroke. The Buda produced 21.7hp on the drawbar and a peak of 30.3 belt pulley horsepower. Fuel economy was measured at 11.4 horsepower hours per gallon. Four forward gears offered top road speed of 10mph. The Model 30 weighed 3,609lb.

◄◄ The Cockshutt Model 30 was the first Canadian-manufactured tractor to be tested at the University of Nebraska. It was sold in the U.S. under license by both Farmers Cooperatives as the Model E-3 and by the Gambles Department Store chain as its Farmcrest Model 30. On each of the U.S. versions, the Cockshutt embossing on the grille was left unpainted but still readily visible. This was the 1949 Gambles.

▲ Raynard Schmidt of Vail, Iowa, restored and owns this Canadian-built row-crop. Cockshutt tractors were among the most stylish of the streamlined machines. The Brantford, Ontario firm (located near Toronto) used an architectural designer to design the machines. Cockshutt was eventually absorbed by Oliver and ultimately became part of the large White Farm Equipment family.

▲The frost of a late fall morning in western Iowa is a long way from the cotton fields of the south or west but as collectors recognize the unique engineering features of certain machines, one region's specialty becomes another's curiosity. The Type 314 low-cylinder cotton picker was mounted on a 1950 Model M. The entire assembly stood 150in tall (basket closed), 104in wide, and 208in long overall. The basket itself measured 72in tall, 104in wide, and 77in long.

◄Bob Pollock of Denison, Iowa, runs the hydraulics on his M to swing open the cotton bin and keep the entire system in working order. International production of the Farmall M ran from 1939 through 1952. In 1940, the tractor (without cotton picker) sold for $1,112. The four-cylinder IHC engine produced 34.4hp on the drawbar and 39.2 on the belt pulley. A three-plow tractor, more than two hundred thousand Model Ms were produced and thousands are still in use.

▸ *The R.H. Sheppard Company of Hanover, Pennsylvania, began producing tractors in 1949 after years of making stationary diesel power units from 5.4 up to 100hp. The SD-3 was its middle of the line machine. The SD-2 and, after 1955, its SD-4, were offered as row-crop or standard tread machines. Early Sheppard tractors had reliability problems with the Detroit Timken Axle Company's rear ends. The SD-4 introduced Sheppard's own rear end, but their reputation was too badly damaged, and Sheppard quit producing tractors at the end of 1958.*

▲Sheppard's engine cylinders had a pre-combustion chamber that allowed very high compression—22:1. Despite this, the Sheppard was a very smooth-running engine. Before International Harvester offered its own diesel for the Farmall M, Sheppard produced a conversion kit. It used the internal pieces but in a block specifically manufactured to mate to the IHC engine mounts and flywheel/clutch assembly. Fuel economy was often quoted as a fifty-gallon drum of diesel plowed forty acres. This 1952 Sheppard is owned and still used by Ray Errett of Harlan, Iowa.

▸Oddly, while Sheppard used an outside vendor rear axle which caused reliability problems and earned a bad reputation, the company made its own fuel-injection system involving a gravity feed right to the plunger. The system had no booster pump. Sheppards were never tested at Nebraska. But the three-cylinder engine had 4.0x5.0in bore and stroke for a total capacity of 188.5ci. At 1,650rpm, the company reported 32 drawbar and pulley horsepower.

▲The Super W-4 was rated as a three-plow tractor. Delivered on rubber tires, it weighed 3,915lb. On-board hydraulics were optional, as was the PTO, which was independent. The International Harvester-built transmission offered five-speeds forward and one reverse.

▲*International's type C-164 engine measured 3.5x4.25in bore and stroke for total capacity of 164ci. Tested at the University of Nebraska in May 1953, the W-4 produced 23.1 hp on the drawbar and a maximum of 31.5hp on the belt pulley at 1,650rpm. Road gear offered a top speed of 15mph. Paul Brecheisen of Helena, Ohio, restored and owns this 1953 Super W-4.*

▶*International Harvester ended the thirteen-year production of its W-4 model in 1953, introducing the Super W-4 at the same time. This was a styled version of the W-4 with a new grille and some changes to the front frame for strength, stiffness, and implement attachment. The tractor measured 82in tall to the top of the stack, 65in wide, and 120in long overall (to the tip of the hitch), on a 67in wheelbase.*

▶▶*The Harris Manufacturing Company of Stockton, California, produced its Model 53 using an industrial Chrysler Model 8A in-line six-cylinder engine. Bore and stroke were 3.43x4.50in for total capacity of 250.6ci. At 2,000rpm, the Chrysler produced a maximum 44.5hp on its drawbar (it had no belt pulley but Harris rated brake horsepower at 62.) The Power Horse steered much like crawler tractors, with hand-operated wheel brakes slowing or stopping either side individually or both together. The Model 53 measured 78in tall, 75in wide, and 107in long. The Laird Welding and Manufacturing Model D-76 blade added 15in to the width and 16in to overall length. This 1954 tractor was owned by the late Tiny Blom of Manilla, Iowa.*

▲Oliver's smallest tractor was produced only during 1957 and 1958. Its manufacture was a kind of company hot potato, beginning at the Battle Creek, Michigan, plant and then moving to the South Bend, Indiana, facilities. Its replacement, the 440, was then produced at the Charles City, Iowa, plant where Hart-Parr tractors were first made.

▶The Super 44 was produced as an offset tractor designed primarily for commercial nurseries and gardens. The offset operating position permitted the tractor's compact dimensions and also allowed the operator greater visibility for cultivating. The tractor was 80in tall to the stack, 54in wide, and 118in overall. The engine was an in-line four-cylinder Continental gasoline, never tested at Nebraska. This 1958 Super 44 was owned by the late Tiny Blom of Manilla, Iowa.

▲*Several—the number is not exactly known—prototypes were sent out for field testing. One tractor was held at the Waterloo factory for work moving equipment and raw materials around the yard. Records indicate the field prototypes were destroyed when they were returned to the factory. One story says the factory yard tractor dragged them off to the cutting torches. This is reported to be the yard prototype.*

◀*This was part of Deere & Co.'s first efforts with articulated four-wheel drive. The Model 8010 was produced only in a small series of prototypes to work out the bugs inherent in something so drastically—and dramatically—different from anything Deere had done before. Bruce Kellar swivels the 8010 in a corn field near home in Kaukauna, Wisconsin.*

▲ It's large. The Model 8010 stands 99in to the top of the steering wheel, 96in wide and—on its 120in wheelbase—it stretches 238in in overall length. Its fuel capacity is 106 gallons. It weighs 26,450lb. Each of its four identical tires is 23.1x26in.

◄ Power for the 8010 came from a six-cylinder two-cycle in-line GMC diesel fitted with a GMC supercharger. Bore and stroke were 4.25x5.00in, with a total capacity of 425ci. The compression ratio of the diesel engine was 17.0:1. Recommended engine operating speed was 2,100rpm which, in eighth gear would provide 18mph on the road. Drawbar pull was established by Deere at 20,000lb, from an estimated 150hp.

▶▶ Deere would go on to produce huge articulated four-wheel drive tractors. However, Walter and Bruce Keller's Model 8010 prototype looks dated alongside a more contemporary Model 4850 equipped with Caster/Action MFWD (mechanical front-wheel drive). A year after the 8010 prototype, Deere introduced its New Generation of Power, with its own in-line four- and six-cylinder diesels. Within a short time, Deere and other manufacturers would add turbochargers and then intercoolers to produce nearly twice the horsepower of GMC's supercharged six.

AMERICA'S CLASSIC
FARM TRACTORS

CLASSIC FARM TRACTORS

RANDY LEFFINGWELL

Acknowledgments

While producing this book, my work interfered with harvesting or planting in many cases. Yet everyone cooperated. This was done not for self-promotion or out of pride of ownership. Rather, this was done to further historical education and promote the hobby of collecting, restoring and demonstrating antique farm machinery. My sincere gratitude goes to the farmers, collectors, and historians who opened their barns, sheds, libraries, and files for this book.

Thanks to Herc Bouris, Sun Valley, California; Harry Case, Red Bluff, California; Virgil Chritton, Pomona, California; Dr. Richard Collison, Carroll, Iowa; Bill Cue, San Diego, California; Cliff Feldkemp, Kaukauna, Wisconsin; Palmer Fossum, Northfield, Minnesota; Dale Gerken, Ft. Dodge, Iowa; Jeff Gravert, Central City, Nebraska; Frank Hansen, Rollingstone, Minnesota; Fred Heidrick, Woodland, California; Joan Hollenitsch, Garden Grove, California; Jim and John Jonas, Wahoo, Nebraska; Kenneth Kass, Dunkerton, Iowa; Walter and Bruce Keller, Brillion, Wisconsin; Lester Larsen, Lincoln, Nebraska; Keith McClung, San Juan Capistrano, California; Mike McGarrity, Pinon Hills, California; Roger Mohr, Vail, Iowa; Ted Nelson, Costa Mesa, California; Robert Pollock, Denison, Iowa; Frank Presley, Gridley, California; Bill Rohr, Compton, California; Gary Spitznogle, Wapello, Iowa; Wes Stoelk, Vail, Iowa; Bob Stroman, Hawthorne, California; and Daniel Zilm, Claremont, Minnesota.

In addition to these individuals, collections were made available for the completion of this book which are open to the public on various occasions. I am grateful to the volunteers at Ardenwood Historical Farm, Freemont, California, for their willingness to reassemble their giant Best steamer. Thanks to the students and faculty advisers of the Agricultural Machinery Collection at University of California at Davis.

To Rod Groenewold, Director, and to the members of the Antique Gas & Steam Engine Museum, at Vista, California, I owe particular thanks.

To my friends Don Hunter and Bill Cox, Pomona, California, and Bob Campbell, Galesburg, Illinois, I cannot express well enough my appreciation for your support, sweat, sense of humor and numerous good ideas.

To Lorry Dunning, historian and educator, Davis, California, I am grateful for the constant challenges and well-reasoned advice. Last and most of all, I thank Virgil White, Sun Valley, California, who labored tirelessly to make this book work.

To all of you who rose early or patiently worked late, who primed cylinders, released compression, spun flywheels or tugged, towed, pushed or shoved tractors into place or introduced me to others who did the same—I dedicate this book.

Randy Leffingwell
Los Angeles

Introduction

The agricultural tractor brought the Industrial Revolution to the farm. But the upheaval cost plenty. The fortunes of the participants—manufacturers and farmers—were the ammunition spent in the war of the machine over horse power.

The history of the farm tractor industry is similar to the development of any hybrid crop. Through trial and error, experimentation and consolidation, some 200 varieties of farm tractor manufactured in the United States in the 1920s were pruned to fewer than a dozen manufacturers selling tractors in the United States today.

Unlike any other revolution in history, this one was dramatically affected by the weather. Enough rain meant profitable crops, which provided the farmer the means to examine and experiment with machine power. Droughts and floods wiped out farms and profits, bankrupting the manufacturers.

The earliest steam technology expanded like the vapor it was. Power increased twenty-fold in steam's forty-year lifetime. Ironically, its greatest surge came in the ten years after gasoline had been introduced and while diesel power was being investigated and developed.

The first portable steam engines were mounted on horse-drawn wagons to transport them. Even when they first propelled themselves, they still resembled horse-drawn wagons since they still used horses to steer.

Previous page
1947 Ford Model 2N Cotton/Cane Special
Palmer Fossum's Sugar Cane/Cotton Special is a bit out of its element in Northfield, Minnesota, corn fields. The Cane/Cotton tractors were manufactured with high clearance and a narrow front end, and meant for the farms and plantations of the south and southeast. The configuration would probably have worked as well for corn.

Early gasoline tractors were similar to the steamers from practicality: they were often assembled from the same parts bins. Gas engines followed steamers' designs because that was what tractors looked like! Early tractor makers were not designers but engineers. It was enough just to get the machines running reliably without also needing to change the appearance.

In the first fifteen years of gas power, tractors grew larger as steamers had done. They were similarly high-priced, because research and development expenses had to be paid off, and also because makers accepted time payments, charging interest in advance. This created a vicious circle: only large farm operators could afford the machinery and they needed equipment large enough to work huge spreads; in 1910, there were 201,000 farms larger than 500 acres.

But early buyers discovered their expensive tractors were useful only for initial prairie sod breaking. They required several people aboard them—just like the earlier steamers—to handle fourteen plow bottoms. Yet these monsters couldn't maneuver between the rows to cultivate. Horses were still needed on the farm.

World War I needed horses—and people—by the millions. The US Army did not yet trust trucks to haul materiel or pull artillery so horses and mules were drafted. The mortality rate was high. The farmers back home were forced to adopt tractors. But what they needed were smaller, one-person machines.

The makers who responded redefined what the tractor could be and could do. Some inventors sold stock to go into business. But some con artists were in business only to sell stock. Many tractors were ill-conceived, inadequately tested, unusable or dangerous. Fortunately for American

1918 Case Model 9-18
Mike McGarrity of Pinion Hills, California, restored and owns this J. I. Case rarity. Many of the initial production was shipped to Europe and South America. Dust and dirt control was important to Case, which incorporated as many moving parts as possible into enclosed castings. Even the huge final drive gears—while not yet fully enclosed in oil baths—were heavily shielded inside the rear wheels.

farmers, one such machine was sold to Wilmot Crozier, a Nebraska legislator.

Crozier purchased a Minneapolis Ford Model B, but barely got it home before it broke down. Then he bought a second-hand Rumely OilPull, which far exceeded its claims. He wondered how many other unreliable machines were out there and if he could force the makers to be more honest.

Crozier proposed his idea to the Nebraska House in 1919. By midyear it was law. To sell a tractor in Nebraska, a firm had to have a sample tested in various prescribed ways by engineers at the University of Nebraska. The tests brought the desired result. Manufacturers who failed either made repairs and tried again, or else they never came back. A Waterloo Boy Model N was the first tractor tested, in April 1920. It passed the first time.

Reliable tractors came from people who grew up on farms and knew from practical experience what was needed. Or they were built by tinkerers. Henry Ford was both. He tested his models for years and then introduced his Fordson, so called because the firm that victimized Crozier also got to Ford: it registered the Ford trademark first.

By adopting mass-production techniques to tractor assembly as he had done to his Model T automobile, Henry Ford produced a tractor less expensively than his competitors. But when production slacked, Ford's labor force and factories cost him money. He cut his price so drastically

that this too cost him money. But even this event redefined the tractor and the business.

At the beginning of the 1920s there were 186 tractor makers in the United States; in the early 1930s there were only thirty-seven. Those that stayed outlasted Ford, who quit in 1928, beaten at his own price war by International Harvester Company.

The International Harvester Farmall—and John Deere's Model D—profited from Ford's innovations. Company owners listened to their engineers who had listened to farmers for years. They introduced a tractor for row-crops, for general farming purposes.

1930 Massey-Harris General Purpose 4WD
Introduced in 1930, it was called the Massey-Harris General Purpose and was rated at 15-22 horsepower. When steel wheels were finally made obsolete after World War II, a number of steel wheel tractor owners cut the rims off and shortened spokes to accommodate rubber tire rims. Or else, they retired the tractor and removed the grousers or else banded them with "road rims."

Meanwhile in California, tractor-makers Daniel Best and Benjamin Holt adapted crawler tracks to their machines. Neither the spongy soil in the river deltas nor the sandy soil around the state could support a wheeled tractor. Holt's Caterpillars changed the US Army's mind about

1928 Deere Model C
The Model C was constructed as a development prototype for Deere's first general-purpose tractor. This Model C, number 200,109, is the oldest known John Deere row-crop tractor, and is owned by Walter and Bruce Keller of Kaukauna, Wisconsin.

mechanization when his crawlers pulled heavy artillery over rough terrain during World War I. Back home, Best's Tracklayers were constantly improved to better suit farmers at home. But the postwar depression caught both. The two merged in 1925 to form Caterpillar Tractor Company.

By the mid-1920s most of the improvements that would see farm tractors through to the 1950s were in place. High-tension magnetos were coupled to spark plugs, and wet-cell batteries replaced the dry cells; self-starters replaced cranks to start the engines.

Dust was the primary killer of tractors. When makers sealed running gear and pressurized lubrication systems, when filters were introduced to clean the air coming into the engines, tractor lives increased tremendously.

Tire companies, anxious to see the entire country on rubber, introduced solid tires for tractors. When pneumatic inflatable tires came along, Harvey Firestone experimented on the tractors at his family farm. Allis-Chalmer's progressive president Harry Merritt hired Indianapolis race-car driver Barney Oldfield to take Allis-Chalmers tractors on inflatable Firestone tires around the United States, racing local farmers at county fairs.

Tractor power was rated from the earliest days by the pull of the drawbar and the power generated on the pulley wheel. It was always quoted with those two numbers in that order: Waterloo Boy's 12-25hp for example. The addition of power takeoff (PTO)—a driveshaft running out the rear of the tractor to power those implements pulled behind it—did not create a new rating. But it made the farm tractor even more flexible.

The worldwide Depression hit the farm hard. Foreclosures were common and became violent. By the mid-1930s, the costs of doing business, the stresses of competition and the Great Depression of 1929 had eliminated nearly half the tractor makers. Only twenty remained—and nine of those controlled virtually the entire tractor market. The number of tractors on farms had more than doubled however, from about 506,000 to nearly 1,175,000 in use.

In 1931, Caterpillar introduced its adaptation of Rudolph Diesel's engine for farm tractor use, in its Model 65. A small gasoline engine used its exhaust to warm the main diesel cylinders and aided in cranking the main engine to life.

Tractor makers observed the auto makers, and tractor designers and engineers learned of outside independents like Henry Dreyfuss and Raymond Loewy, the "streamliners" of America. Dreyfuss and Loewy simplified manufacturing and improved farmer safety on these sleeker machines. Each maker introduced separate logo colors.

In 1939, Henry Ford returned to the business with a remarkable new device and a new name attached to his tractors. Irishman Harry Ferguson had invented a three-point hitch that allowed lightweight tractors to work like heavyweights. The patented Ferguson System was so significant that every tractor-maker mimicked it for the next fifteen years until the patents expired.

World War II further refined American tractor production. The need for steel and rubber removed thousands of tons of obsolete tractors and machinery—future collectibles—from farms. The War Production Board advertised: "An old tractor can yield 580 .30 calibre machine guns." The board reduced manufacture of farm equipment nearly one third but increased spare parts production by half. Quotas were established.

By 1945, the number of tractors on farms had doubled again since the mid-1930s to nearly 2,422,000. Progressive farmers were fascinated by tractors but economic realities held them back. Horses ate whether working or not, but they were paid for and their food was apportioned from the total farm—usually about one fifth of total acreage. Tractors required fuel and repairs. Both had to be paid for in cash in town. The choice was tough.

1920 Avery Model C
The 1920 Model C used Avery's six-cylinder engine, which produced 8.7 drawbar and 14.0 belt pulley horsepower at 1250rpm.

In 1947, James Cockshutt, a Canadian equipment maker, introduced the continuous-running independent power takeoff. Before this innovation, the PTO shaft stopped running when the tractor drive clutch was disengaged, stopping the implement as well. Cockshutt fitted a separate clutch for its PTO; work continued whether the tractor was crawling at 1/2mph, running at 4mph or standing dead still.

A surge of new developments flooded the early 1950s. Improved hitching systems sped up implement attachment. Transmissions and gear-reduction systems—some providing eighteen speeds—let tractors get maximum engine torque at the ground speed the farmer chose. Liquified propane gas (LPG) was introduced as an inexpensive alternative to high-octane gasoline. Power steering joined power implement lift to give the farmer's arms and back a rest. Even tractor seats got much-needed attention. Enclosed cabs insulated and protected farmers from noise and climate, making long days seem somewhat shorter. Rollover risk was met with mandatory adaptation of tall roll bars.

Deere, long the traditionalist, threw itself into the 1960s with a new series of four- and six-cylinder engines. For many, this New Generation of Power signalled the sad end of an era; for others, it marked the first day they considered buying a John Deere.

Four-wheel-drive, produced with little success in the 1920s and 1930s, returned in the late 1950s. By the early 1960s, as the American manufacturers spiraled up horsepower competition again, four-wheel-drive became the most efficient way to transmit 150hp, 250hp, and more to the ground.

By the late 1960s, tractor history began repeating itself. Power output had increased so much that small tractors (less than 25hp) were uncommon in American manufacturer lines. But the Japanese and the Germans—whose average farm sizes were much smaller than those in the

1921 Samson Model M
The Samson Model M used Samson's own four-cylinder L-head engine, which produced 11.5 drawbar and 19.0 pulley horsepower.

United States—had them. West Germany's Deutz joined the US market first with air-cooled diesels in 1966. Satoh, part of Mitsubishi Agricultural Machinery Company, was among the first Japanese makes in the US market, bringing in a 22hp four-cylinder gas tractor in 1969.

By 1972, US manufacturers each offered large articulated four-wheel-drive tractors. At first, all the wheels steered but the body did not bend; fronts turned, rears turned, or front *and* rears turned, and crabbing was possible. With Minneapolis-Moline's 1969 A4T, the body bent in the middle. A complex kind of universal joint allowed the halves to twist as well, accommodating extreme changes in terrain.

Dual wheels front and rear denoted the next major spurt in power. Specialty manufacturers like Steiger, Rite, Rome, and Versatile offered turbocharged diesels with 800ci and 900ci producing 400hp to 600hp. These weren't show tractors meant for competition; these were workers, pulling disc harrow gangs 60ft wide across miles of wide-open fields.

1936 John Deere BO and 1939 International Farmall F-20
Following the introduction of Deere & Company's 9–16hp rated Model B, several variations were offered, including industrials and orchard models. Orchard tractors lowered the operator's seating position and cleared the hood of any stacks and pipes that could damage the trees. This 1936 BO, left, belongs to Bob Pollock while the 1939 Model F-20 International Harvester Farmall belongs to his father, Raymond. Purchased in 1948, the tractor is operated every day and has only missed work long enough to have its valves ground once.

Caterpillar stretched the technology envelope with its rubber-tracked Challengers. Steel reinforcement held the tread together even when run on paved roads to the fields. The 31,000lb machine exerted half the ground pressure of comparable wheeled tractors. These 325hp diesels came standard with a dozen floodlights.

The Vietnam War lingered. Oil-producing nations in the Mideast formed the Organization of Petroleum Exporting Countries (OPEC) to control output and increase revenues. Fuel prices jumped, launching another recession. Labor disputes crippled companies not already struggling. US unemployment approached 10 percent.

The family farm was hit from all directions. Trade protectionism crippled overseas produce sales, further stockpiling unwanted grain. Prices fell. Farmers again faced foreclosure. In 1940, one in four Americans lived on the farm; by 1980, only one in twenty remained.

Burdened with equipment and education debts, many more farmers sold out. Corporations, looking to diversify and searching for tax losses—something farmers knew well—bought farms. Foreign corporations with low-value US dollars acquired potentially high-value rural real estate. Foreign investment in farm land exceeded $1 billion by 1980.

The giant companies sought their comfort in mergers. Harry Ferguson, after divorcing Ford, joined Massey-Harris in 1953. In 1969, Oliver, Cockshutt, and Minneapolis-Moline merged under the umbrella of White Farm Equipment. In the 1980s, companies with funds that could have developed new machinery, or reduced costs, instead spent that money fighting off hostile takeovers.

In 1984, International Harvester joined J. I. Case, part of Tenneco since 1967. In 1985, Deutz of West Germany bought Allis-Chalmers. Ford acquired New Holland of Pennsylvania, moved to England and joined with Italy's Fiat. Caterpillar, which introduced its Challengers, and John Deere, which brought out its 6000 and 7000 Series turbocharged diesel tractors in 1992, emerged intact, bruised but still independent.

New tractor development suggests more power with more features. But it also hints at a vibrant life for antique tractors. Each year the hobbies of restoring and operating these machines draw a more knowledgeable and more involved audience.

Chapter 1
Allis-Chalmers

Allis-Chalmers, Ford, John Deere, White, and J. I. Case: names of companies and names of men. But these companies were run and ruled by their namesakes, men of larger impact. These were men whose names stayed on the front door a century after they were gone.

Allis-Chalmers grew by acquiring and consolidating the innovations of others: James Decker and Charles Seville; John Nichols and David Shepard; Cornelius Aultman and Henry Taylor; Meinrad Rumely; and Edward Allis.

Metal work and machinery were the common background. Financial successes and failures brought them together.

Decker and Seville's Reliance Works flour milling company survived only fourteen years before bankruptcy forced the sheriff's auction where Edward Allis was the high-bidder. By January, 1868, Allis' Reliance had expanded. The first Allis steam engine came from its new works near Milwaukee, Wisconsin.

Around this time, John Nichols in Battle Creek, Michigan, produced his first thresher. With David Shepard as a partner, his company achieved renown for his vibrator separator. It engaged Henry Taylor to promote its machine.

1928 Allis-Chalmers Model 20-35
Allis-Chalmers introduced the Model 20-35 in 1921. With this model, A-C introduced its air washer. An exhaust-pipe-high (hidden behind the exhaust in this view) air-intake pipe fed air through a cast-iron water bath. Some reports suggest this was also cooling water—in order to make freezing weather operation possible—but definite information is unavailable. The 20-35 weighed 7,095lb and at 930rpm was capable of 3.25mph in the higher of its two forward gears. Introduced originally at $1,885, by 1928, in order to compete, Allis-Chalmers had reduced the price to $1,295. By then, the pinstriping, swivel seats, and even axle hub covers were extra-cost options. This example is owned by Fred Heidrick of Woodland, California.

However, in Canton, Ohio, Cornelius Aultman was also manufacturing separators. Taylor—traveling for Nichols & Shepard—met Aultman. Taylor changed employers and continued his job as travelling salesman, for Aultman's company.

Meanwhile a German emigrant, Meinrad Rumely, opened a blacksmith shop in La Porte, Indiana. Meinrad and brother John produced their first thresher, and in 1859 beat thirteen others in a competition at the United States Fair. Rumely's first stationary steam engine came in 1861.

Aultman & Company had incorporated in 1865. Henry Taylor bought some Nichols & Shepard's patents and manufacturing rights and in 1867, Aultman, Taylor & Company was set up in Mansfield, Ohio. Aultman & Taylor Machinery Company was established around 1893, producing threshers and steam traction engines.

Aultman made its first steam traction engines in 1889 with horizontal or vertical boilers, burning straw, wood, or coal. While it introduced its first gasoline engines in 1910, Aultman offered steam through the mid-1920s, selling nearly 5,900 steamers.

1915 Rumely OilPull Model M
Owner Bill Rohr of Compton, California, explains to some friends the workings of his 1915 Rumely OilPull Model M. The two-cylinder 6.812x8.250in engine produced 27.5 drawbar horsepower when tested at the University of Nebraska. The 8,750lb tractor sold new for nearly $2,000.

The Financial Panic of 1873 caught Edward Allis overextended and bankruptcy hit the Wisconsin operations. His own reputation saved him and reorganization came quickly, as the Edward P. Allis Company. He set out to hire known experts: George Hinkley, who perfected the band saw; William Gray, who revolutionized the flour-milling process through roller milling; and Edwin Reynolds, who ran the Corliss Steam Engine works.

When Allis died in 1889, Hinkley, Gray, and Reynolds continued innovation and expansion. By 1900, Allis was the largest steam engine builder in America. A new 100 acre Wisconsin plant site, at West Allis, opened in September 1902.

Then William Chalmers, president of a machinery and stamping mill firm, met Edwin Reynolds. E. P. Allis Co. was successful and Reynolds believed Allis could control the industrial-engine business. Chalmers' company was failing and he saw in his firm an opportunity to provide Allis with additional plant capacity.

On May 8, 1901, Allis-Chalmers Company was incorporated, immediately capable of supplying much of America's manufacturing needs. But financial troubles arose again. So in early 1912, a former Wisconsin National Guard Brigadier General, Otto Falk, was drafted to take over the company renamed: Allis-Chalmers Manufacturing.

Over in La Porte, Meinrad Rumely's stationary engines led to portable and traction engines, with locomotive-style boilers. Ever curious, Rumely heard about John Secor's experiments. In 1885, Secor tested low-grade distillate fuels in internal-combustion engines; he developed a kerosene-burning ship engine in late 1888. Meinrad Rumely died on March 31, 1904, and was succeeded by his two sons, William, as president, and Joseph, as secretary and treasurer. While Joseph's son, Edward, was studying medicine in Germany, he met Dr. Rudolph Diesel. Diesel's first engines "burned" coal dust but eventually he too worked with low-grade fuel oil.

Rumely asked Secor to produce a practical tractor engine. Within ten months, Secor and a staff of designers completed a prototype tractor. Production began in early

1918 Rumely OilPull Model 14-28
The Model 14-28 was introduced in 1917, as Rumely's entry into the "small tractor" market. Weighing 9,600lb, this 1918 Model 14-28 sold for $2,400. In 1911, M. Rumely Company purchased the Advance Thresher Company; financial difficulties soon forced two reorganizations, the first in 1913. The later problems, caused by Rumely's rapid expansion and farmers' crop failures, gave birth to the Advance Rumely Company.

1910. While workers referred to Secor's prototype as "Kerosene Annie," Edward Rumely and his secretary named it the OilPull. The first production model was completed in February and by year end, 100 were completed.

In 1911, Rumely bought Advance Thresher Company of Battle Creek, Michigan, established in 1881. In a short-lived partnership with Minneapolis Threshing Machine Company (MTM), they produced friction belt-driven steam traction engines from 1886 and sold something like 12,000 steam engines. MTM took its share of proceeds and eventually became part of Minneapolis-Moline.

Rumely's 1912 production was 2,656 tractors. Staff swelled to 2,000 employees. "Kerosene Annie" went into production as the Model B OilPull, rated at 25-45hp at 375rpm. It weighed a hefty 12 tons. Rumely bought Northwest Thresher Company in late 1912 and picked up Northwest's 24-40 tractor. Rumely reclassified the tractor as a 15-30hp machine and named it the GasPull, and produced it at the former Northwest shops until 1915.

But 1913 was still ahead. And 1914 was worse yet. The 1913 sales were only 858 OilPulls, roughly one third of 1912's boom year. A disastrous crop failure in Canada made banks nervous; they refused to extend credit. Dr. Edward Rumely resigned January 1, 1914, but things got worse. In 1914 only 357 OilPulls sold. In January 1915, M. Rumely Co. filed bankruptcy. The family lost its fortune and the firm as well.

Once under receivership, new tractor development resumed. The 1918 Model 15-30 introduced a high-tension magneto and optional spark plugs, which increased its power rating to 18-35; a 14-28 was also introduced. But these were short-lived: the 18-35 was discontinued by 1920, the 14-28 was improved to rate 16-30hp.

Aultman & Taylor introduced its first gas tractor late in 1910. The 12 ton 30-60hp machine used a four-cylinder

Previous page
1918 Rumely OilPull Model 14-28
The two-cylinder horizontal engine measured 7.00x8.125in bore and stroke and produced 14 drawbar and 28 pulley horsepower at 530rpm. Standard ignition was a low-tension magneto with mechanical igniters. But a high-tension system was offered and could be installed either before the initial purchase or afterwards by the operator himself.

1918 Rumely OilPull Model 14-28
In 1918, operators wrote away for parts. The mail took weeks for the round trip. Now, obsolete parts brokers accept orders by telephone or fax and ship by overnight express when necessary—and if they have it. Otherwise the owner makes his own. That path is the one Mike McGarrity followed most often when restoring number 9872.

1928 Allis-Chalmers Model 20-35
The Model 20-35 used Allis-Chalmers' upright four-cylinder engine with 4.75x6.50in bore and stroke. At 930rpm, 23.6 drawbar and 38.6 pulley horsepower was produced. Ignition was provided by an Eisemann Model G-4 magneto. A "hot-rod" version, the Model E 20-35, was introduced mid-1928, with Eisemann's GS-4 magneto contributing to a higher output: 33.2 drawbar and 44.3 pulley horsepower.

horizontal transverse-mounted engine. Self-starting was prevalent by now. The 30-60 used an on-board high-pressure air tank to turn over the engine. Magneto spark was provided by storage dry-cell batteries to start.

Aultman & Taylor suffered difficulties; on January 1, 1924, Advance-Rumely took over. While inventories of

parts and tractors remained, Advance-Rumely marketed Aultman & Taylor's machinery. By September, Aultman & Taylor's name disappeared.

When General Otto Falk arrived at Allis-Chalmers as court-appointed trustee, he declared war on the marketplace. He launched a small tractor, a "tractor-truck" (akin to

1928 Allis-Chalmers Model 20-35
It was another two years before Harry Merritt traveled to California and witnessed the hillsides covered with wild spring poppies. His fascination with the sight and his desire that all A-C tractors stand out on their own hillsides led to his ordering all tractors after 1930 to be painted Persian Orange, the closest color he could find to California poppies.

1928 Allis-Chalmers Model 20-35
The Model 20-35 appeared in several variations through the years. Some changes were no doubt cost-cutting moves. Beginning in 1927, for example, rear fenders that had previously covered the rear wheels down to the operator's floorboards were cut back to the "short-fender" style.

the military half-track trucks to appear later), and then a tractor-tiller. Falk's tractor truck appeared as a long-wheel-base flat-bed truck with crawler treads at the rear beneath the bed. It was meant to draw plows and carry loads. But its price was $5,000, roughly seventeen times the price of Henry Ford's Model A truck.

The tractor-tiller tricycle apparatus featured a wide row of tines on a rotating axle dragged behind the machine. It too disappeared quickly.

Falk's machines were curiously innovative: the first tractor was built on a one-piece heat-treated steel frame. The second was conceived as a full-system machine, with its engine driving two large wheels in front. The implement-of-choice became the rear-wheel assembly. Its resemblance to Moline Plow Company's Universal prompted a warning from Moline about patent infringement.

But Falk had succeeded four years earlier. In December 1919, Allis-Chalmers produced a stylish 15-30hp tractor.

1931 Allis-Chalmers Model U
Sitting on the paved test track at the Lincoln campus of the University of Nebraska, this 1931 Model U, number U25-1, is a replica built by A-C to commemorate the Model U's significance as the first tractor offered for sale on pneumatic rubber tires in the United States. Three years after the U, in May 1934, A-C brought its new Model WC as the first rubber-tired example ever tested. Rubber tires immediately increased drawbar pull 15 percent and fuel economy 25 percent over comparable steel-wheeled competitors.

1941 Allis-Chalmers Model M
A-C bought Monarch tractors in 1928 and continued producing Monarch's existing models until 1931. At that time, A-C submitted a new Model K to the University of Nebraska for testing. Model L and M came immediately after. The Model M—such as this 1941 example owned by Herc Bouris at Minifee Valley, California—remained in production until 1942.

Next page
1941 Allis-Chalmers Model M
Crawlers are simple machines with enough pedals and levers to keep an operator busy. A main clutch, gear shift, and hand throttle break up the symmetry of the track clutches and track brake pedals. A-C's Model M has four forward speeds with a top speed of 5.8 mph. The crawler weighs 6,620lb but could pull nearly 5,200lb in low gear at 2.15mph.

Its looks were inspired by automotive designs. A-C reclassified it in February 1920 as the 18-30. Bad timing slowed the 18-30 to a halt shortly after introduction.

The US economy had harvested a depression after World War 1. Sluggish sales of the Fordson led Henry Ford to cut prices by more than 50 percent. Those who could compete with Ford dropped their prices too. But others who only thought they could compete went broke. Allis-Chalmers sold only 235 tractors in 1920.

Previous page
1941 Allis-Chalmers Model M
The inline upright four-cylinder engine measures 4.50x5.00in bore and stroke and produced 29.65 drawbar horsepower in its Nebraska Tests. Ignition was provided by an Eisemann GL4 magneto and A-C used a Zenith K5 carburetor below the large Vortex air-cleaner.

1948 Allis-Chalmers Model G
Introduced in 1948, Allis-Chalmers' Model G was meant to appeal to farmers, gardeners, and even highway departments. Its unusual rear-engine configuration provided exceptional visibility for users of one-, two-, or even three-row cultivators. This version, equipped with the rare optional dual wheels, is owned by Erik Groscup of Escondido, California.

Allis introduced a smaller tractor, the 12-20 in 1921. But it was another victim of the Depression and the tractor price wars; only 1,705 sold by 1928. Yet life resumed for the survivors when the economy loosened up. Tractor production was twenty times as great for Allis in 1928. Total sales neared 16,000.

For Advance-Rumely too, the Ford-International Harvester tractor wars had been near fatal. But in October 1924, Rumely introduced the new lightweight OilPulls. The 15-25L weighed 6,000lb and introduced a Rumely patent, the locking differential. In addition, all the gears in the L were completely enclosed and the transmission ran in ball-bearings. But very few "lightweights" were produced.

Advance-Rumely had acquired the Toro Motor Cultivator rights in 1927. This led to the Do-All. Introduced in 1929 as a convertible tractor/cultivator combination, this was Advance-Rumely's first *true* lightweight. It attempted to meet the growing demand for "all-purpose" tractors.

In mid-1931, Advance-Rumely was absorbed into Allis-Chalmers, which now became the fourth-largest farm equipment maker in the United States. When Advance-Rumely inventories disappeared, their tractors went out of production.

General Falk took a beating in the first half of his twenty years with A-C. Force-feeding tractors to the company cost Allis money. But Falk persevered. The tractor department became well-established and in 1926 Falk brought in another bright talent, Harry Merritt.

Merritt was a progressive innovator. The tractor department was Falk's favorite so Merritt received encouragement too. In 1929, a new tractor was prepared by Allis for an outside marketing firm, United Tractor & Equipment in Chicago. A three-plow rated machine, the United tractor used a four-cylinder Continental engine. When United went under, Allis took back the tractor, called the Model U.

Merritt had traveled through the West. In California in the spring, he fell for the bright orange wild poppies covering the hillsides. Returning to Milwaukee, he reexamined A-C's somber green tractors. "Persian Orange" most closely

1948 Allis-Chalmers Model G
A-C fitted a four-cylinder Continental engine, Model AN62, to its Model G tractors. Tested at Nebraska in 1948, the 2.375x3.50in bore and stroke engine produced 7.3 drawbar and 10.1 belt horsepower at 1800rpm. Its top speed in fourth was 7mph.

matched the wild California hillsides. It wasn't long before Merritt changed their color.

A-C sold more than 10,000 Persian Orange Model Us through 1944. Continental engines powered the early models until A-C's own UM four-cylinder appeared in 1933. The U was offered in a variety of styles, including an "Ind-U-strial" model, a crawler model, a row-crop version, and railroad yard switcher built by Brookville Locomotive.

But it was not longevity, adaptability or even its new color for which Allis-Chalmer's Model U was most famous. It was for Harry Merritt's friendship with Harvey Firestone.

Firestone's family farm, the Homestead, ran on A-Cs on steel wheels. Harry Merritt offered Firestone a new Model U. Firestone, devoted to pneumatic tires on the road, wondered about their use on the farm. The same economy of operation, operator comfort and reduced vibration would apply...if pneumatic rubber tires had adequate traction.

Airplane tires were mounted on truck rims first. But Firestone understood the tread needed to grip in earth, sand, wet clay, sod. This gave birth to the connected-bar design—the continuation of one side of a chevron to the bar above it.

On Firestone's Homestead Farm, every tractor and implement was refitted with his new "Ground Grip" pneumatic tires. Merritt changed the specifications on the Model U. Inflatable rubber was offered as an option. Merritt's brightly colored Model U became the first tractor in the United States offered with rubber. Allis' next generation row-crop, its WC introduced in 1934, was the first tractor designed with inflatable rubber specified as standard equipment.

The success of Merritt and Falk's tractors was also due to full-line implements available for each machine. Mechanical farming was possible for every crop. Virtually every size farm was served after the 1937 introduction of their 1,900lb, $570 Model B. Their 1940 Model C introduced distillate fuel engines. Rated as a two-plow tractor, it introduced Allis' Quick-Hitch system for rapid attachment and release of cultivators and other equipment.

A new system of implements and tractors was introduced in 1948. The four-cylinder rear-engined Model G was Allis' smallest to date. The Model WD introduced the "Traction Booster," A-C's version of Ferguson's three-point hydraulic hitch; the WD also utilized two clutches—the foot clutch disengaged all engine power whereas the hand

1948 Allis-Chalmers Model G
Without the cultivator tool bar, the side-mounted sickle bar, and the dual wheels, the standard Model G weighed 1,549lb. Its standard tires are 4.00x12in fronts and 6x30in rears. Without accessories, the Model G sold for $970.

clutch merely stopped tractor motion. Lastly, it offered power-adjust rear track. WD's were sold in single, narrow and standard front ends through 1953, when Allis introduced its Snap-Coupler system, a more versatile replacement to the quick-hitch. A more powerful replacement, the WD-45, offered LPG models from the start. It was the first tractor with factory installed power-steering.

It was not until 1955 that Allis-Chalmers had a diesel in its line. But in 1957 an entire new line of tractors cleared the boards, introducing new features. A "Roll-shift" front axle used the power steering system to change front track much like its rear axles. A "Power Director" added a high-range to each transmission gear.

Sales success continued until 1980. The agricultural equipment market began to shrink. In March 1985, Allis-Chalmers Agricultural Equipment Group was sold to a subsidiary of Klockner-Humboldt-Deutz AG of West Germany. Deutz-Allis was born. Persian Orange was resprayed Deutz spring green.

Then in April 1990, the original Allis-Chalmers Agricultural Equipment Group was reacquired by a group of American investors. By July, new American management—while retaining Deutz-Allis' name—reinstated Allis' signature corporate Orange on all domestic manufactured machines. The poppies were back.

Chapter 2

Case

Jerome Increase Case was 23 years old when he moved to Wisconsin, taking with him six Groundhog threshers purchased on credit. He settled near Racine. Born in New York, he grew up assisting his father operating and selling threshers. That experience helped Jerome sell his first five. He kept the sixth, to work with and earn money, and later it became his test bed for his own ideas. His work evolved into a harvester that combined threshing with cleaning that separated the wheat completely from the chaff. By late 1844, he had perfected his new harvester and was producing it in Racine.

In 1863, his twentieth year in business, he incorporated J. I. Case & Company. At the end of the Civil War, he adopted the company logo, "Old Abe," the eagle. It was Wisconsin's state bird and had been the battlefield emblem of the Eighth Wisconsin Regiment from Eau Claire.

Case continually experimented and improved its machines, regularly adopting ideas from its competitors. In 1869, it introduced a portable steam engine to replace horse-powered treadmills and rotary sweeps. Case's reputation for quality and innovation continued with the steamer and it sold well for decades. But it wasn't until 1884 that Case brought out his first steam traction engine. The direct-flue-type boiler provided motion but horses were still needed to steer it. Soon after, a steering wheel was attached to a worm gear and chains to pull the front axle left or right. Case ultimately produced steam traction engines as large as 150hp and as late as 1924. Only nine of the giant 150 hp machines were produced from 1905 through 1907. Each weighed 18 tons dry, and sold for $4,000.

J. I. Case died December 22, 1891. The next year, the company's first gas-engine tractor was tested. The desire for smaller tractors continued even as Case's steam traction engines proliferated; a total of 35,838 were sold by the end of steamer production in 1924. The peak production year—2,322 tractors—was 1911. By that time, however, Case had solved some initial problems—of fuel mix and spark—and it introduced the 30-60. The introductory version won first place at the 1911 Winnipeg Tractor trials.

1948 Case Model VA
Announced in late 1939, The Case Model V was the one- or two-plow rated little brother to the big four-plow LA. Then in 1942, the VA replaced the V, replacing the V's Continental four-cylinder engine with Case's own four. The Model VA row-crop version was called the VAC and it was introduced at the same time as the VA standard. These were not tested at University of Nebraska until October 1949 (testing was interrupted between 1942 and 1946 because of World War II). Both distillate and gasoline versions were tested and the gasoline performance—15.0 drawbar and 20.1 belt horsepower—was predictably slightly better than the distillate at 12.5hp and 17.0hp. This 1948 VA is part of the collection of the Antique Gas and Steam Engine Museum in Vista, California.

1915 Heider Model C
The 1915 Heider Model C was built in Carroll, Iowa, by John Heider, a farmer-turned-tractor maker. Rock Island Plow Company acted as sales agent with so much success that it eventually bought out Heider. Rock Island produced the Iowan's tractors until 1927. The Model C used a Waukesha four-cylinder 4.50x6.75in engine rated at 12-20hp.

The 30-60 continued in production through 1916. The tractor, with its horizontal two-cylinder engine, sold for $2,500. A "compact" version, introduced in 1913, rated 12-25hp and sold for $1,350. And a 20-40hp mid-range tractor was also introduced, winning two gold medals at the 1913 Winnipeg trials. Its opposed two-cylinder engine used engine exhaust to induce air current over the radiator. This system was replaced by a water pump and fan.

In the mid-1910s, farmers were intrigued by three-wheelers. Case responded with its 10-20 in 1915. More significant was its first use of a vertical-mount four-cylinder engine. The engine, fitted transversely across the cast frame, was fully enclosed. Case then introduced its compact 9-18 in 1916.

The next cross-mount cast-frame Case was the 15-27, introduced at the end of 1918. The engine differed by being cast with all four cylinders "en bloc," in the same casting. It breathed through an early water-bath air cleaner. The largest of Case's lineup arrived at the beginning of 1920 and was rated a 22-40hp tractor. It was similar to the 9-18 and 15-27 with its crosswise-mounted four-cylinder. Dissimilar, however, was the frame. Case reverted to channel iron built up and assembled rather than the one-piece castings on the smaller machines.

But constant improvement and upgrading was Case's way of business. Its 10-20 tricycle was replaced by the new 12-20 in 1922, using a one-piece cast frame that placed the engine, transmission, and rear axle in one rigid assembly. The engine cylinder head, transmission cover, and locking differential cover bolted on to provide additional rigidity, yet allow for easier serviceability. The transverse-mounted engine drove through bull gears and pinions, sealed inside the frame. Dirt was eliminated and the drive gearing was simplified. Its compact design—9ft 1in long, a 24ft turning circle, and 4,230lb weight would still pull three 14in plows.

It sold for $1,095, and like many tractors of the day, was meant to start on gasoline and switch over to kerosene or other distillate.

Engine speeds increased and cylinder-head designs improved for better fuel and exhaust flow. But the days of small, heavy, and relatively expensive tractors were numbered. In 1922, J. I. Case commissioned a study of the market support for a new type of tractor. The company continued producing its cross-mount standards through the 1920s but in 1929 it unveiled its new L, still a standard configuration but with the engine now mounted lengthwise. The L had been tested and developed to the highest level of efficiency. A scaled-down version, the C, was introduced soon after and was offered in orchard and industrial versions from the start.

Case had other business affairs occupying its financial attentions: the other J. I. Case company, the Plow Works, owned Wallis Tractors. In 1928, Case Plow Works was sold to Massey-Harris in Canada who wanted the Wallis and Case plows but not Case's name. J. I. Case Threshing Machine Company bought that back for $700,000, ending years of confusion between two separate companies founded by the same man, in business with such similar names.

In August 1928, Case bought Emerson-Brantingham, the Rockford, Illinois, tractor and implement company.

1915 Heider Model C
Heider's tractor used a friction drive in which the entire engine slid forward or backward to speed up or reverse tractor direction. The tractor weighs nearly 6,000lb and sold new for $1,100. It is now owned by the Carroll, Iowa, Historical Society and was restored by members and volunteers.

Emerson had introduced its "foot lift" plow in 1895. Through the decades, E-B acquired a number of other companies, producing implements, tools, and tractors, among them the Gasoline Traction Company of Minneapolis, and its Big 4 tractors.

Nearly a decade later, in 1937, J. I. Case purchased Illinois plow and implements maker, Rock Island Plow Company, for its tillage and harvesting equipment. In 1914, Rock Island had agreed with Heider Manufacturing of Carroll, Iowa, to market its tractor. Heider's friction-drive tractors resembled the earliest Case friction tractors; Heider's own engine moved forward or backwards to engage drive and increase speed. Rock Island Plow continued manufacturing Heider tractors until 1927.

In 1932, Case brought out the row-crop CC, advertising it as "2 tractors in 1," boasting adjustable rear wheel spacing from 48-84in, power-lift to raise and lower the implements, independent rear wheel brakes, and a 5.1mph top speed. PTO was optional as was a standard front wheel axle.

1918 Case Model 9-18
Introduced in the fall of 1916, Case's Model 9-18 was produced only for about fifteen months, until spring 1918. Responding to the cries from magazine and newspaper editors for compact tractors, Case mounted its four-cylinder engine sideways on the frame, shortening the wheelbase substantially and beginning a technology that the J. I. Case Threshing Machine Company would continue until the 1929 introduction of the Model L.

In 1935, Case introduced its "motor-lift," which raised implements on the fly with the touch of a button. A simplified mounting system for its new implements made equipment change faster. Case's all-purpose tractor was available on rubber.

The R Series was introduced in 1936, using a Waukesha 3.25x4.00in four. It was available as a standard, all-purpose, or industrial tractor until 1940. With the R tractors, styling arrived at Case. The new radiator grilles reflected a spray of wheat.

1918 Case Model 9-18
Case produced nearly 6,500 of these Crossmount 9-18 tractors, the first half of production in late 1916 and through 1917. Beginning in 1918, the company introduced the 9-18 Model B, which replaced Case's usual structural steel frame with an all-cast version, to further increase strength and decrease weight. The 3,200 built in 1918 were all Model B versions.

Previous page
1918 Case Model 9-18
Air intakes were available in two versions for the 9-18. One version fitted a 2ft stack to raise the air intake above the dust layer. The version shown here picked up air at the radiator top and filtered it by centrifugal force, dumping the sediment into the ubiquitous Ball glass jar. From there, the outside cool air traveled by a tube surrounding the exhaust pipe to warm it before it got to the carburetor.

1918 Case Model 9-18
J. I. Case's four-cylinder measured 3.875x5.00in bore and stroke and rated 9 drawbar and 18 pulley horsepower at 900rpm. In addition to frame structure changes, the company increased engine speed for the Model B versions in 1918, rating peak power at 1050rpm. This did raise drawbar power to 10hp and led to the introduction later in 1918 of the Model 10-18.

1922 Case Model 12-20 Crossmount
The cast-frame technology continued with the introduction in late 1921 of the compact Model 12-20, to replace the tricycle Model 10-20. A new four-cylinder cast-en-bloc engine measured 4.125x5.00in bore and stroke, ran at 1050rpm and produced 13.2 drawbar and 22.5 belt horsepower. It weighed 4,450lb, and is now owned by Fred Heidrick of Woodland, California.

Case's new Model D was shown under spotlights to dealers at the Racine head office in 1939. Gone was the gray paint scheme that twenty years earlier had replaced the cross-motors' green. "Flambeau Red" joined the palette of other colors plowing, cultivating, and harvesting US farm fields. Flambeau, the French word for torch, was also the name of a Wisconsin river. Recalling its Wisconsin Civil War regiment's eagle mascot, Case introduced "Eagle-Eye visibility" the result of a higher, fully adjustable operator seat and more vertical steering wheel.

The row-crop D's four-cylinder engine with the four-speed transmission was good for 10mph on the roads. The

1925 Case Model 25-45
J. I. Case had underrated the performance of its Model 22-40 introduced in 1920. As a result, when it was tested again in 1924 at the University of Nebraska, it bore a new designation, the Model 25-45. The performance at the late fall test far exceeded even its new rating. The Model 25-45 was manufactured through 1929 but was not renamed.

1925 Case Model 25-45
Case inaugurated its cross-mounted engine design with its 1918 Model 9-18 tractor and by 1925, a full lineup followed. Case offered tractors from as small as 12-20hp-rated 4,230lb machines up to massive 21,200lb machines rated at 40-72hp. The configuration was encouraged by public demand for "small" tractors, which Case took to mean small in length if not in gross horsepower.

next year, Case replaced the R with its new V in all versions. A two-plow rated S Series came out as well. The VA replaced the V in 1942, and introduced Case's first high-crop tractor, the VAH. The company continued the S and VA series as well as the Ds in production until 1955.

In 1953, Case changed its numbers and colors, and added diesel power. A 500 Series six-cylinder diesel pro-

duced 56hp, standard with dual disc brakes, electric lights, and starter. The 400s came in 1955, and the 300s arrived in 1956. The 500s brought two-tone paint: Flambeau Red with Desert Sand.

The new series Powr-Torq engines ran on gas, LP-gas, distillates, or diesel. Tripl-Range twelve-speed transmissions provided 12mph in road gear. The Eagle Hitch three-point implement hook-up monitored plow depth and increased traction. The 300s rated three plows, the 400s four plows, and in 1957 the 600s offered six speeds to pull six plows. In 1958, Case introduced 700, 800, and 900 Series tractors

1925 Case Model 25-45
Fred Heidrick's 1925 Model 25-45 was photographed near the end of its restoration process but still missing its tall air cleaner intake pipe as well as the full exhaust pipe. A variety of rear drive-wheel configurations were offered, these being standard steel lugs with a 6in, extension rim. Road rims offered hard rubber grousers and rice field wheels fitted lugs 36in long, extending a full 18in beyond the rim.

with Case-O-Matic transmissions. The 900 diesels with six-cylinder 377ci engines rated 70hp.

In 1957, Case entered the crawler market, purchasing American Tractor Company of Churubusco, Indiana—

Previous page
1925 Case Model 25-45
Case used a Kingston Model L carburetor as well as the American Bosch ZR-4 magneto to fuel and fire its vertical-but-transverse-mounted four-cylinder engine. Each cylinder measured 5.50x6.75in. At 850rpm, the engine produced 32.9 drawbar horsepower and 52.6 belt pulley horsepower—impressive for a tractor largely unchanged from its original 22-40hp rating.

1948 Case Model LA
J. I. Case began painting its tractors Flambeau Red during the summer of 1939 so all Model LA tractors appeared in the new corporate colors. Powered by Case's own 4.625x6.00in four-cylinder engine, its 1952 Nebraska Tests reported 41.6 drawbar and 55.6 belt horsepower at 1150rpm. Cliff Feldkemp of Kaukauna, Wisconsin, restored and owns this example.

makers since 1950 of the Terratrac. An innovative company, American Tractors' GT-25 and GT-30 gasoline and DT-34 Diesel offered interchangeable track gauges for row-crop applications, a three-point hydraulic hitch, and track shoes of either steel or rubber. Through the 1950s, Case continued Terratrac production at Churubusco.

In 1967, J. I. Case was purchased by Kern County Land Company, which was then purchased by Tenneco, a Houston, Texas, conglomerate founded on oil. Two years later, the Case eagle, Old Abe, retired. In 1972, Case acquired David Brown Tractors, of England.

In the next ten years, the economy began shrinking from the effects of war in Southeast Asia and from indus-

1948 Case Model LA
Introduced in 1940, Case's Model LA was the company's big tractor for more than a dozen years. Electric starting was a $73 option and rubber tires raised the tractor price by nearly $300 to around $1,650. By 1952 when this version was manufactured, the price had risen to $3,000.

trial overproduction. Mid-Eastern oil producing nations raised gasoline and diesel fuel prices while food prices dropped. In 1980, the Federal Reserve Bank raised the prime lending rate to 21.5 percent. Bank financing dried up and tractor sales all but ceased.

Federal farm policies supported high grain prices at a time of immense surpluses. Overseas shipments plummeted. Correspondingly, total tractor sales of all colors were only 29,247 in 1981, less than some single-year single-model sales in the past. Dealers went under by the hundreds.

1948 Case Model VA
The V Series tractors offered a top speed in transport gear of 10mph. Case offered VAO orchard versions as well as an industrial VAI. Many of these were sold to the US armed forces and used to move materiel around supply depots and jet aircraft around Air Force bases. The V Series remained in production until 1955.

Previous page
1948 Case Model VA
The Model V used Continental's 3.00x4.25in long-stroke four-cylinder engine rated to 1425rpm. When the VA was introduced in late 1942, Case was building its own engine for the tractor. With bore and stroke of 3.25x3.75in, it was still slightly oversquare, which still produced good torque. The electric starter versions sold for $742 at the factory in Racine, Wisconsin.

1948 Case Model VA
The VAC weighs 3,200lb and provided adjustable rear tread width from 48in to 88in to accommodate all kinds of crop rows. An optional hydraulic system was available to incorporate the three-point Eagle Hitch, Case's version of the Ferguson three-point system. Case charged $205 for this option, over the standard equipment price of $1,388.

1951 Case Model LA
Beginning in 1952, Case offered Liquid Propane Gas as an optional fuel source for its LA. The company aggressively promoted this option, emphasizing its economy of operation compared to its power output. Use of kerosene and distillate fuels had ended because their power output was often nearly 20 percent less than gasoline.

Tenneco considered selling J. I. Case because its losses were so serious. When International Harvester joined in 1979, the situation became acute but in 1988, Tenneco decided instead to sell its oil company, for $7.5 billion. After that, Case-IH as it was now known, became the largest division of the restructured Tenneco.

1951 Case Model LA
To accommodate Liquid Propane Gas (LPG), Case increased the compression ratio on its four cylinder 4.625x6.00in engine from 5.75:1 for gasoline to 7.58:1. This resulted in comparable performance between the two. Gasoline engines produced 41.6 drawbar and 55.6 belt horsepower at 1150rpm while the LPG engines produced 40.9 drawbar and 57.9 belt horsepower.

1951 Case Model LA
The LPG Model LA weighed nearly 7,788lb, the extra tank, carburetor, and regulator adding nearly 170lb to the weight of the gasoline versions. Nebraska Tests ranked maximum pull from the LPG version at 6,874lb at 2.27mph whereas the gasoline version pulled 6,659lb at 2.26mph.

1951 Case Model LA
This big Case was sold on 15x30 rear tires and 7.50x18 fronts. By 1948, when this tractor, number 5200085, was manufactured, the Model LA on pneumatic rubber tires sold for around $2,120. When Liquid Propane Gas was offered as a factory option in 1952, it added $179 to the price. This model is on display at the Antique Gas and Steam Engine Museum in Vista, California.

1951 Case Model LA

1951 Case Model LA
This Model LA was manufactured in 1948 originally as a gasoline-powered tractor. But when LPG became an option, it was retrofitted with the tank, regulator, carburetor, and intake manifold to run the cheaper, more-efficient burning Liquid Propane. Case built its own engine.

Chapter 3

Caterpillar

The discovery of gold in Sutter's Creek near Sacramento in 1848 moved more people to California than any other event in history. Most of those who moved were farmers. And when the Sutter's Creek gold harvest was completed, most of the farmers became Californians and discovered the other gold of California. These were the fields down the center of the state where the gold was planted—2 1/2 million acres of wheat. Harvesting this new gold proved more lucrative than Sutter's Creek. Newspapers at the time advertised harvester rates up to $3 per acre, meaning $7 million in 1890.

For Case, McCormick, and other harvester makers back East, the only route to California for their machines was by ship around South America. Yet, faced with vast acreage, wealthy farmers ordered machinery from back East or the innovative farmers built their own.

Daniel Best was 26 years old when he moved from Ohio to Oregon in 1859 to hunt for gold. After little success, he moved to California in 1869 to join his three brothers farming north of Sacramento. When he had to haul grain into town *and* pay $3 per ton for cleaning, he decided to "bring the cleaner to the grain."

In 1870, Best built three cleaning machines. Business flourished and Best manufactured to his factory's capacity. Watching customers fit cleaners to its combined harvesters, Best modified its own machine. By 1888, it had produced 150 Best combines.

Meanwhile, Benjamin and Charles Holt, of C. H. Holt, a lumber and wagon materials company already in business nineteen years, were making wagons at Holt's subsidiary, Stockton Wheel Company. Holt had expanded to Stockton in 1883, seeking a drier climate for aging wagon wood.

Holt's machines, unlike its competition, used "link belts"—chains with replaceable links easily reassembled in the field—to drive all the moving parts. It pioneered the use of V belts, tapered leather belts, to connect threshing cylinder shafts to the drive.

By 1900, Holt had produced 1,072 combines, more than all its competition together. It conceived a "side hill" harvester with adjustable wheel height and header angle so

1926 Caterpillar Model 60
In 1925, after the two financially ailing companies agreed to "merge" for a second time, Best and Holt tractors all fell under the name "Caterpillar"—even while most of those in the new lineup were formerly Best machines. An example is C. L. Best's Model 60, which by 1926 when this example was manufactured, was renamed the Caterpillar 60. Best's—and Caterpillar's—Model 60 was powered by a four-cylinder valve-in-head engine of 6.50x8.50in bore and stroke. In Nebraska Tractor Tests, the Best version in 1924 produced 72.51 gross horsepower—equivalent to pulley rating— at 650rpm. The 60 weighed 20,000lb, not including the LeTourneau bulldozer blade and winch attached.

1902 Best 110hp Steamer
Daniel Best's steam traction engines produced power in proportion to match their size. This 110hp model—one of 364 built—stands 17ft, 4in tall, 28ft long, and rolls on 8ft, 2in tall rear drive wheels. Its immense power could haul 36tons of rolling load up 12 percent grades. Restored and owned by Ardenwood Historical Farm, Freemont, California, volunteer members occasionally fire it up.

that the harvester box was always vertical on any hillside. Still, driving the harvesters with horse or mule teams was not ideal. Animals had to be fed, watered, harnessed, or turned out. And they still made only slightly more than 1mph. Holt and Best both believed better power was available.

The steam traction engine had first appeared in the United States as the product of Obed Hussey and Joseph Fawkes in Pennsylvania in the mid-1850s. The technology arrived in California in the 1860s, built by Philander Standish and Riley Doan. By 1886, the first steam-powered combine operated in California, burning straw due to the scarcity of wood and coal. De Lafayette Remington, of the

firearms family, claimed he'd made the first steam traction engine on the West Coast in Oregon in 1885, and patented in 1888. Best saw it and bought the patents, which he quickly improved.

Best's boiler stood upright between the rear drive wheels on a heavy frame. The two main drive wheels were 8ft tall

1909 Holt Model 45
While most of Ben Holt's crawlers used a single front "tiller wheel," those meant for the Midwest used two front steering wheels. Pliny Holt, Benjamin Holt's nephew, was head of the Northern Holt Company located in Minneapolis. Plans called for assembly of ten such machines but only two were completed. This rare machine belongs to Fred Heidrick of Woodland, California.

1913 Best Model 30 "Humpback"
Tractor-makers Best and Holt knew their markets. Californians raised wheat on vast expanses of sandy soil or they raised fruit grown on trees. Specialized equipment was needed. Best's "Humpback" 30 was a forerunner of the orchard tractor. By placing the final drive gears above independent track idler wheels, Best lowered the entire profile more than 1ft. The full canopy roof and high exhaust suggest this model never worked orchards, however.

and 30in wide. The drive was geared and applied to the rim of the wheel, not the axle. The machine weighed 22,000lb and produced 60hp at 150lb of steam pressure.

By 1889, Holt had produced its first steamer, nicknamed "Betsy." Friction arms operated the flywheels and the engine was reversible but there was no transmission or differential. Betsy weighed a trim 24 tons but was reported able to pull thirty plow bottoms and plant 40 acres a day.

Holt and Best competed vigorously with similar machines. But Holt concentrated on steam-powered combined harvesters where Best specialized in traction engines. By the time of their merger in 1925, Best had built 1,351 steam combines, while Holt had built 8,000.

1917 Holt Model 75
The 1917 Holt Model 75 sold new for about $5,500, its substantial price including interest in advance on the time payments. Driven by Holt's 1400ci four-cylinder gasoline engine, each cylinder measured 7.50x8.00in, the 23,600lb machine produced 75hp at 550rpm. Don Hunter of Pomona, California, restored and owns this example, and runs it frequently.

Their machines worked best for planting and harvesting of grain. But they were expensive, nearly $5,000; harvesters cost another $2,500. Only the largest farms could afford steam power.

Holt's business grew. It was already aware of the farming value of the river bottom land in the San Joaquin Valley. And it knew of tractors on tracks. In 1903, Ben Holt and his nephew, Pliny, toured Europe and the United States to see what had been done by competitors. Holt harvesters and tricycle steamers bogged down in the river deltas where the richest soil could not even support a horse. So in late 1904, Holt adapted its 40hp Junior Road engine to rear tracks. Each track measured 9ft long and 2ft wide, of 2x4in boards attached to link chains so that drive sprockets moved the chains.

On Thanksgiving Day, November 24, 1904, the steam crawler was first tested. Ben Holt watched it with a friend, Charles Clements, who was hypnotized by the motion of the track undulating across its upper guide rollers. He said it crawled like a caterpillar. Holt took to the name immediately.

Competition between Ben Holt and Dan Best led to a lawsuit charging infringement on Best's patents by Holt. Holt lost the first trial in December 1907, but on appeal in the summer of 1908, it pointed out that every significant feature of Best's harvester had appeared on another inventor's machines. In August the ruling was overturned and returned to lower court for retrial. But on October 8, 1908, Best sold its business to Holt for $325,000. Both firms

1917 Holt Model 120
By a trick of the camera lens, Fred Heidrick's massive 1917 Holt Model 70-120 is dwarfed by freshly turned clods of northern California soil. Holt made its international reputation with these 120hp crawlers. Most were exported to Europe to tow artillery during World War I. For perspective, this tractor is 21ft long, 10ft tall. Each of its six cylinders measures 7.50x8.00in.

knew the end of steam power was coming. They combined assets, dealers, and technologies, operating under the Holt name.

Experiments with gasoline engines impressed Ben Holt, and led to successful tests in December 1906. Manufacture began in 1907; the first four tractors were sold in 1908. Number four—the first agricultural sale—went to a nearby farmer.

But the first three joined other Holt steam tractors working on the Los Angeles Aqueduct project—a $23 million project to bring water from central California's Sierra Nevada mountains to Los Angeles through a concrete "river" built across the Mojave Desert. Holt's 100hp steamers hauled 30 tons up a steady 14 percent grade. Orders followed for another twenty-eight tractors over six months.

The desert mountains offered different challenges from the soft San Joaquin River deltas Holt knew. Dust and heat

1919 Holt Midget 18hp
Dwarfed by its bigger cousins, this 1918 Holt Midget sits in front of Fred Heidrick's collection of future projects, several Best and Holt crawlers in 60hp and 75hp variations. The "Midget" is only 4ft 5in tall and wide, barely 15ft long and weighs only 6,920lb. Engines on the 75s weighed nearly as much.

caused such frequent and expensive repairs—in time out of service and in material costs—that parts of the project reverted to mules.

During the aqueduct project, Pliny began producing tractors in Minneapolis in 1909. Pieces for ten 45hp gasoline tracklayer tractors were shipped; two were completed. These were "designed" for the Midwest with two front steering wheels. Crawlers, necessary in soft Western soil, were less valuable in Midwest soils but cost $1,000 more than their competitors.

Pliny Holt met an implement dealer from Peoria, Illinois, who knew of a large plant whose owner had bank-

1919 Holt Midget 18hp
The tiny red arrow rising from the steering gear was really necessary to indicate to the operator which way the tiller wheel was headed. Seated low, some 15ft away behind a long hood, the operator could not see the wheel. Holt's track clutches and brakes provided most of the steering. The front wheel became a front-mounted plow if not for the arrow to show its direction.

1919 Holt Midget 18hp
Half of its four-cylinder engine shown here, the Midget produced 18hp on the pulley, 8hp on the drawbar. With three speeds forward, the compact crawler was good for nearly 4mph flat out. With standard 11in tracks, Holt advertised the tractor as the perfect machine for work in narrow vineyards or citrus orchards.

rupted recently. On October 25, 1909, Holt acquired the Peoria facility. And when Holt incorporated in Illinois the following January, a new name was born: The Holt Caterpillar Company.

Best's son, C. L. "Leo," had stayed with Holt only two years before moving to Elmhurst, California, a short distance from San Leandro. His new company, C. L. Best Gas Traction Company, began immediately to produce wheeled and crawler tractors.

Holt again found itself in competition with a Best. The rivalry centered again around service, quality, and design. Dan Best had offered 50hp and 110hp traction engines in 1900. By 1910, his son offered a 60hp gas tractor. A year later, the 60hp was replaced by an 80hp. Leo's first crawler was a 75hp introduced in 1912.

1919 Holt Midget 18hp
Introduced in 1914, Holt advertised its Midget as "The smallest size built, it has the guaranteed power of eight horses...for the sort of work that no other tractor and no other size Caterpillar can handle. Light ground pressure adapts it to work on soft ground—the '18' won't mire, nor slip, nor pack the ground." New, they sold for $1,600.

1923 Best Model 30
C. L. Best built its own engines, drivetrains, frames, and bodywork, relying on such outside suppliers as Bosch for magnetos, Ensign for carburetors, and Pomona for air cleaners. Best's inline four-cylinder engine measured 4.75x6.50in bore and stroke. As tested at Nebraska, the Model 30 weighed 8,100lb and in low gear, pulled 4,930lb at just 2mph. This 1923 Model 30 is owned by the Antique Gas and Steam Engine Museum of Vista, California.

Best's tractors were painted black with gold lettering while Ben Holt's were brown with yellow trim. Best's were called "Tracklayers." Holt registered Clements' term, "Caterpillar," as a trademark in 1910. And so the rivalry continued.

Best's Tracklayers were similar to Holt's Caterpillars. Best's system used a differential while Holt's drove each side separately, eliminating the differential. Best had a huge differential, which enabled them to pull as hard around a corner as on the straightaway. Both tractors were steered by power and had brakes for each track. Best tractors could not turn quite so sharply as Holts, probably owing to Best's differential continuing to move the inside track.

From 1914–1918, the war in Europe affected Best and Holt dramatically—and differently. Holt shipped Caterpillars to the front—virtually all its production was taken by the US Army—while Best shipped Tracklayers around the United States. The agricultural market was served by C. L. Best, whose dealer network numbered fifteen by 1918.

1923 Best Model 30
By 1921, other improvements appeared on C. L. Best's crawlers, some resembling Holt's improvements: the master clutch to the engine and drivetrain was now supplemented by separate track clutches—steering clutches—to improve maneuverability. Rated as a 20-30hp tractor, in Nebraska Tractor Tests, a similar machine produced 25.96hp on the drawbar at 800rpm in low gear.

Next page
1923 Best Model 30
By 1923, Best—and Holt—had concluded front tiller wheels were no longer necessary. Best's Model 30, still offered as a humpback, was also available in standard crawler configuration. Sometimes known as the "Muley," the 30hp models were first introduced in 1914 and sold for around $2,400.

After the war, Best broadened its line. By 1925, it had forty-three dealers and nearly twice the business it had five years earlier. Best tractors were well-developed and had profited from farmer input. Its flexible tracks oscillated over small rocks and uneven ground without pitching the tracklayer from side to side.

Holt's military contract reflected its aggressive marketing. Before the war, the Army remained convinced only of the mule's ability to pull supply trains. Bad experiences led them to rule out trucks for materiel supply. Holt offered to demonstrate its tractor anywhere at its own cost. By the time the Army took a look in May 1915, Holt had already sold tractors to France, Russia, and Great Britain.

Holt's tests succeeded. When the Army took bids on tractors, only Holt bid. When those first twenty-seven were delivered, Holt already had sold more than 1,200 to the Allies. Holt initially shipped its 75hp Caterpillars to Europe. However, in 1917 it produced a new high-powered 120hp six-cylinder. An erroneous newspaper account even credited Holt with the invention of the military tank after its tractor sales to the English.

Holt expanded tremendously during the war, employing 2,100 workers. The US Army ordered a total 1,800 45hp tractors, 1,500 75hp tractors, and ninety of the big 120s. Holt's war output was substantial, a total of 5,072, with nearly 2,100 to the Allies.

1926 Caterpillar Model 60
Despite its not quite show-piece appearance, this nearly 68-year-old workhorse still starts on the first tug on the flywheel. The plate on the flywheel warns the operator to shift gears only when the tractor is standing still. The three forwards speed transmission shift pattern is marked. Engraved in large type is the advice to move the gear shift lever *without force*.

1926 Caterpillar Model 60
Owner Virgil White of Sun Valley, California, estimates the LeTourneau 8ft bulldozer blade and logging winch apparatus at the rear add easily another 6,000lb to his 1926 Caterpillar 60. White threatens to restore the tractor but its reliability and strength have made it the most relied on workhorse at the Antique Gas and Steam Engine Museum of Vista, California, where it lives protected from the elements only by its rust.

But this left no time to expand domestic sales, no personnel to service it, and a limited line of products due to the US government specifications. Holt came out of four years of heavy production with large inventories, tractors ill-suited to agricultural needs, and poor cash flow. Peace treaties canceled all the wartime contracts.

By 1920, the United States was in a nationwide depression. High wartime production levels overtook reduced postwar demand. Holt was caught from both sides: American farmers who needed machinery to meet wartime food demands lost those markets in peacetime; worse yet, the government, oversupplied with tractors, flooded the market with "war surplus."

When Ben Holt died in December 1920, his successor, business manager Thomas Baxter, cut the large tractors from the lineup, introducing smaller models more suited to agricultural purposes. He learned of a $1 billion federal highway building fund and began directing company advertising to road contractors.

Yet Best was in no better shape. Leo Best had fought Holt's marketing trick for trick. But the fierce competition

cost each company severely. Combining and regrouping resources, Best and Holt again "merged" on March 2, 1925, into a new corporation, The Caterpillar Tractor Company.

Shortcomings in Holt's lineup were also consolidated. Best's 30hp and 60hp tractors survived and were renamed

1929 Caterpillar Model 15
In 1928, Caterpillar introduced its compact Model 15, selling for $1,900 new. The little 6,000lb crawler continued in production through 1932. This restored example is part of Fred Heidrick's extensive collection of Best, Holt, and Caterpillar tractors.

1929 Caterpillar Model 15
The 15 is powered by Caterpillar's own inline vertical four-cylinder L-head engine, of 3.75x5.00in bore and stroke. Tested at the University of Nebraska, the 15 produced a maximum of 21.3 drawbar horsepower and tugged a load of 4,166lb in low gear at 1250rpm. Caterpillar used an Eisemann G4 magneto, Ensign carburetor, and the Pomona canister air cleaner.

Caterpillars. Best's forty-three US dealers merged with its seven exporters. Total sales jumped up 70 percent even as Ford and International entered price wars to sell tractors.

Holt brought a worldwide reputation, plant facilities six times the size, and sales revenues still twice the size of Best's. The gasoline four-cylinder engines, Caterpillar's mainstay, were put on final notice. A new engine, converted from gasoline, was out testing the theories of Dr. Rudolph Diesel. In 1931, seventy-five of the Series 60 diesels were produced.

In 1935, Caterpillars changed designations, eliminating tonnage displacement or horsepower nomenclature. Rudolph Diesel's initials were paired with a number to de-

1929 Caterpillar Model 15
Prior to the merger of Best and Holt in March 1925, Holt's Caterpillar lineup designated its sizes by the tractor weight. Holt marketed tractors called Two-Ton, Five-Ton, and Ten-Ton. These continued in production after the merger until stocks were exhausted. Thereafter, models were referred to by conservative rating of horsepower; the Two-Ton was replaced by the Model 15 and Holt's Ten-Ton was replaced by the Caterpillar Model 60.

1932 Caterpillar Model 60
Used for maintenance work on the Panama Canal after its opening, this Model 60 is now owned by Virgil Chritton of Pomona, California. Six such Model 60 tractors were used for more than a decade on road construction and shoreline renovation. These machines then found their way to southern California and played a role in the development of Orange County.

scribe the tractor's size. The RD6, RD7, and RD8 were followed by the RD4 in 1936. Standard gasoline-engine tractors were simply labeled the R Series. The designations were simplified further—letter D for diesel alone—in 1937 and the D2 followed the long Caterpillar line.

The agricultural industry in the West was peculiar in the United States. While all the other manufacturers responded to Eastern and Midwestern needs, developing smaller tractors of greater versatility, the Western farmer with his vast acreage, needed big-tractor power.

1934 Caterpillar Model RD-4
Caterpillar's experiments with diesel power led to the introduction of the RD line, so named to honor Dr. Rudolph Diesel, the father of the technology. Introduced in 1934, the RD-4 was roughly the equivalent of the Caterpillar 30, which ran on gasoline or distillate fuels. In later versions, the designation became simply D-4. This example with the low operator's position is owned by Virgil White of Sun Valley, California.

1934 Caterpillar Model RD-4
The RD-4 was tested at the University of Nebraska in October 1936. It weighed 10,000lb, and used Caterpillar's inline four-cylinder engine with 4.25x5.50in bore and stroke. At 1400rpm the engine produced a maximum of 39.8hp. This early version started with a two-cylinder gasoline engine that served as power to turn over the diesel crank. To start the gasoline engine, operators used a lawnmower-type rope pull.

Caterpillar's experience and financial achievements led many agricultural competitors to think again about crawlers. But agricultural tractor history is riddled with tales of makers awakening to the need for general-purpose tractors and then trying to become all things to all farmers. When Caterpillar briefly experimented with a "Midwest" tractor, it had failed. But Cat didn't change the tractor for its new market. It encouraged a new use, in construction, and created a new market for its old tractor.

1935 Caterpillar Model 22
Caterpillar Model 22 tractors ran on 10in tracks and would operate on either gasoline or distillate fuel. Overall, the tractor was 8ft long, about 58in wide, and 56in tall, and could turn around inside a 10ft circle. It sold new for $1,450 at the factory.

1935 Caterpillar Model 22
The Model 22 was introduced in 1934 and used Caterpillar's inline four-cylinder engine of 4.00x5.00in bore and stroke. Tested at the University of Nebraska, the gasoline engine produced 19.3 drawbar and 27.2 belt horsepower at 1250rpm, with a maximum pull of 4,900lb in low gear. A test with distillate fuel produced a 4,534lb pull. The tractor weighed nearly 7,400lb.

1935 Caterpillar Model 22
In the mid-1930s, Caterpillar nomenclature became complicated. Some models adopted RD and R prefixes while others retained the number series begun a decade earlier. All the number series, such as this Model 22, were gasoline engines. RD models were diesels and produced more power from the more efficient fuel. R crawlers were gasoline only, and very few were produced. This 22 is owned by Keith McClung of San Juan Capistrano, California.

Chapter 4

Deere

The gumbo soil near the Rock River at Grand Detour, Illinois, was much stickier than in John Deere's native Vermont. It required constant scraping from iron plows. Deere, a blacksmith, had an idea. Starting with a broken steel sawmill blade, highly polished by thousands of cuts, he cut off the teeth, and formed and fitted it to a wrought iron and wood handle. A farmer tried it and after the steel plow sliced through the soil cleanly, he ordered two.

Deere, who had arrived in Grand Detour at age 33 with—the story goes—$73.73 in his pocket, went into a new business. A partnership soon led Deere and his factory to Moline, on the Mississippi River. By October 1843, the new factory had finished its first ten plows. Deere's son Charles joined in 1853. Charles enjoyed getting out and meeting the customers and demonstrating the equipment, though his training was in keeping the books balanced and bills paid.

Charles opened branch offices for quicker response to customers and cash flow. By 1900, Deere & Company sold cultivators, harrows, seed drills and planters, wagons, and more. "New Deal" plow combinations up to six gangs were shown in catalogs. While horse teams had pulled two plows, something more powerful was obviously in mind at Deere. In 1889, Deere's brochure showed the New Deal six gang plows pulled by a steam traction engine.

William Butterworth, a lawyer and Deere's next president, followed tradition. But mechanization was coming to the farm. In 1908, the Winnipeg Agricultural Motor Competitions began, showing both steam traction and gasoline engine tractors. By 1909, Deere plows were represented and in 1910, the Gas Traction Company of Minnesota's Big Four 30hp won the gold medal pulling a seven-bottom Deere plow.

The machine was in Deere's 1911 catalogs. Deere saw the need more than ever for a tractor of its own. Engineer C. H. Melvin was assigned to design and build a prototype in June.

His tricycle prototype was complete in 1912 but it failed field tests and work stopped by 1914. The board persisted. Board member Joseph Dain was asked to work on a small tractor. Dain's first efforts appeared in 1915, following the style of the day for "lightweight" tricycle tractors.

1928 Deere Model C (Left) and 1928 Model GP-WT (Above)
John Deere built the Model C, left, as a development prototype for a general-purpose tractor. Farmers asked for a tractor to do cultivation as well as first plowing and harvesting. Deere built 112 of these over three years, then recalled them when they introduced the Model GP in 1928. A few escaped, including this, number 200,109, the oldest known John Deere row-crop tractor. It is owned by Walter and Bruce Keller of Kaukauna, Wisconsin. Above, as with the Model Cs, most were called back and later sold as updated production models. At 900rpm and running on kerosene, the tractor engine produced 17.2 drawbar and 25.0 belt pulley horsepower. Only two of these are still known to exist. This one, number 204,213, is also owned by Walter and Bruce Keller of Kaukauna, Wisconsin.

311

1915 Waterloo Boy Model R
The Waterloo Gasoline Engine Company began producing tractors called the Waterloo Boy in 1912. The model styles and engineering subtly evolved through and beyond this 1915 Model R Style D, among the last versions sold with a vertical fuel tank. This Waterloo Boy was restored and is owned by Kenneth Kass of Dunkerton, Iowa.

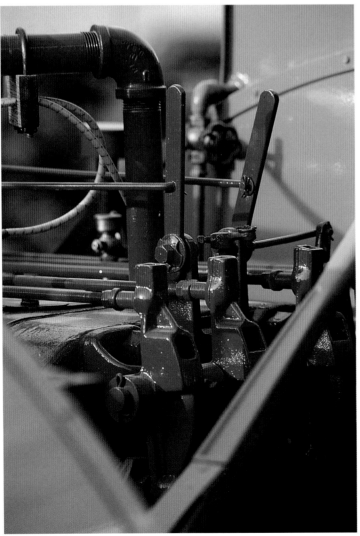

Dain's machine was all-wheel-drive, with chains propelling the front and rear axle. It worked well enough to merit additional development. Ten more prototypes were built using a more powerful engine.

Dain's all-wheel-drive John Deere tractor would sell for $1,200. It was approved, but one Deere executive, concerned about the time from manufacturing start-up to first sales, suggested instead that Deere & Company should buy an existing company. The company making Waterloo Boy tractors was available.

John Froelich and two partners formed Waterloo Gasoline Traction Engine Company in 1893 to produce gasoline engine tractors based on his experiments. But after some failures, Froelich left. His former partners renamed the company the Waterloo Gasoline Engine Company, and

1915 Waterloo Boy Model R
Waterloo Model R tractors used the firm's own horizontally mounted side-by-side two-cylinder engine. Bore and stroke measured 5.50x7.00in, and power was rated at 12 drawbar and 24 belt pulley horsepower. Waterloo used Schebler carburetors and a Dixie Model 44 low-tension magneto. Peak power was reached at 750rpm.

1915 Waterloo Boy Model R
Only 115 models of the first 5.50x7.00in engine versions were built by Waterloo Gasoline Engine Company up until mid-1915. Afterwards, engine displacement increased—the bore by 0.50in—which raised belt pulley output from 24hp to 25hp. The Model R had only one speed forward and reverse, for a maximum of 2.5mph. It weighed more than 6,000lb.

concentrated on stationary engines for pump and sawmill power. In 1906, the name Waterloo Boy was adopted, playing on the title of the "water" boy required for steam tractors but not needed with gas engines. By 1912, new engineers had developed several tractors and had built twenty-nine by 1914; nine were constructed as "California Specials," or tricycles. The other twenty were four-wheeled tractors.

The two-cylinder engine for a new model, the R, was completed and about 116 were then sold. Waterloo introduced an improved engine for a new Model N. The R had sold for $985; the N with a two-speed transmission sold for $1150. They all produced a distinctive "pop pop" exhaust sound.

A report to Deere's Board explained the advantages of Waterloo's two-cylinders over competitors' fours: economy

1918 Dain All-Wheel Drive
The 79th tractor to bear a John Deere farm tractor name and number plate was this John Deere All-Wheel Drive, designed for Deere & Company by board member Joseph Dain. Tested through 1915 and 1916, regular production began in 1917 with 100 of these authorized from the newly established tractor division. This rare piece of history is owned by Frank Hansen of Rollingstone, Minnesota.

1918 Dain All-Wheel Drive
Cooling was a problem with the original development models of Deere & Company's All-Wheel Drive tractor. Finally, to cool the big engine, a large Perfex brass-cased radiator was fitted. By the time of its introduction, Deere engineers had eliminated all the problems from the All-Wheel Drive tractor—except for factory capacity to build it.

of construction, reduced number of parts needed to build or repair, accessibility of repair areas in a horizontally mounted engine, and fuel economy.

The $2,350,000 purchase was approved. Overnight Deere was a tractor maker. Other projects followed, involving "cultivators," tractors meant for mid-season work. The cultivators set the stage for "general-purpose" tricycle-wheeled tractors. And these row-crop machines influenced tractor design forever.

Need for a four-cylinder engine was debated—Ford and IHC both had them. Money ruled: it would cost too much to design, machine, test, and produce a new engine. The

1918 Dain All-Wheel Drive
Originally, Joe Dain's design specification called for a Waukesha four-cylinder engine. However, problems with cooling and inadequate power sent Deere & Company to William McVicker to design a powerplant just for Deere. With 4.50x5.00in bore and stroke, the cast-en-bloc detachable-head engine developed 12 drawbar and 24 belt pulley horsepower.

1926 Deere Model D
Designed by the Waterloo Gasoline Engine Company even before Deere & Company purchased its factory and grounds in March 1918, Deere inherited the plans and development test models for a new, compact tractor known as the Model D. By the time this 1926 model was manufactured, Deere had abandoned spoke flywheels for solid flywheels. Herc Bouris of Sun City, California, owns this Deere and the 1931 Ford Model A dump truck in the background.

new John Deere Model D would use the two-cylinder engine. Introduced early in 1923, it would sell for $1,000.

The first Ds placed the steering on the left, directly linked to the front axle. After 879 were built during 1923 and 1924, the company changed from a 26in spoked flywheel to a thicker 24in flywheel of equal weight. A jointed steering rod, seen on experimental tractors, was fitted into the 4,876 production machines in 1924 and 1925.

For 1926, a solid flywheel replaced spokes. In 1927, Deere enlarged engine bore. Exports began that year: forty-six to Argentina and Russia. The peak export year, 1929, Deere shipped 2,194 to Argentina, 2,232 to Russia. In 1930, 4,181 were delivered to Russia. For 1931, steering moved to the right side, with a worm and gear system. Governed engine speed rose. An industrial Model D was in-

1926 Deere Model D
The Model D used Deere's horizontal two-cylinder engine, adopted from the Waterloo Boy tractors. The 1926 Model D was a three-plow-rated tractor, with bore and stroke of 6.50x7.00in. Peak power in tests at the University of Nebraska, was 22.5 drawbar and 30.4 belt pulley horsepower, reached at 800rpm from the engines made famous by their distinctive "Johnny Popper" exhaust sound.

1926 Deere Model D
The Deere Model D came into being as a result of pressure on all the tractor makers from Henry Ford, who had introduced the Fordson in 1917. Deere's compact Model D reduced its tractor dimensions from 132in to 109in in overall length and from nearly 6,200lb in late-model Waterloo Boy models to just 4,000lb in the early Ds.

troduced in 1926—hard rubber tires front and rear and high-speed gears allowed a top speed of 5mph.

The Model D standard-tread designs pointed up the difficulties of mechanizing row-crop agriculture. Deere set out to design the row-crop all-purpose tractor in 1925. By July 1926, the first three Deere Model C prototypes were ready as two-bottom plow tractors.

The Model C's immediate triumph was a mechanism using engine power to lift the cultivator from the ground. This innovation quickly appeared on all the competition. The C was first to provide four forms of power—the lift, a power takeoff, drive belt, and draw bar. But dealers worried

1926 Deere Model D
The operator's position remained on the left side for the earliest versions of Model D production, switching to the right to improve steering control and accuracy beginning with the 1930 models. Deere continually improved and evolved the Model D, and kept it in production through 1953.

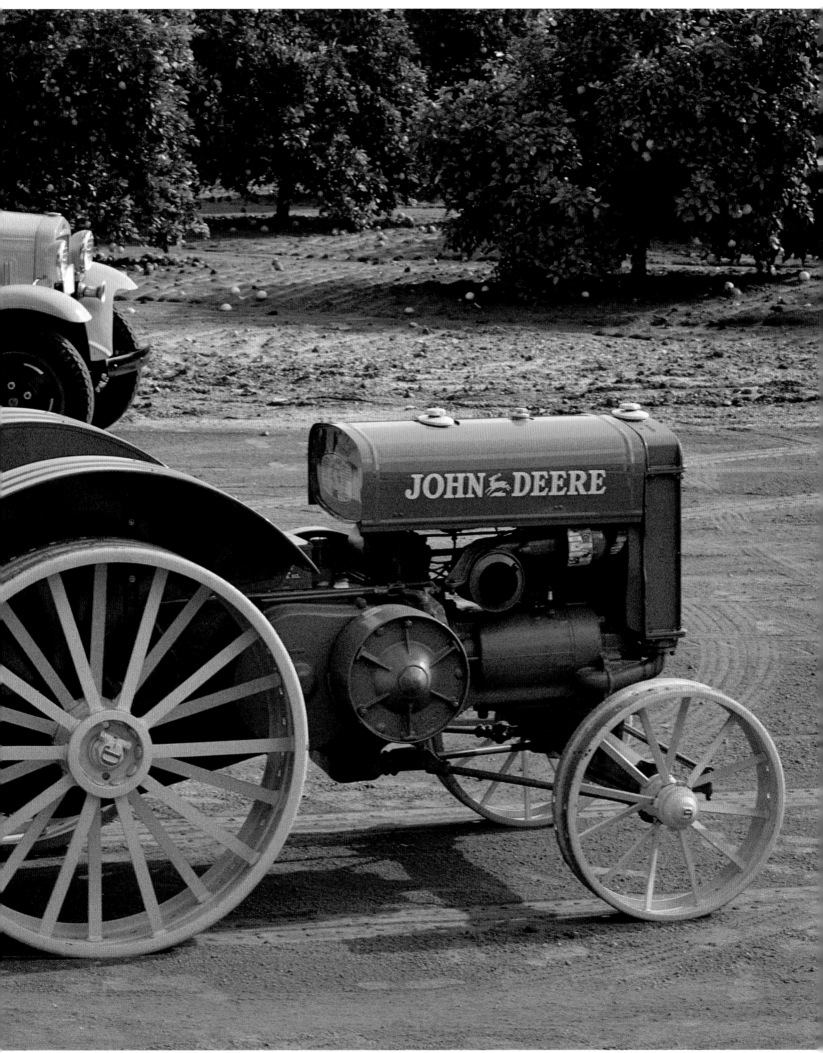

Previous page
1928 Deere Model GP-WT
The Model C served as the development tractor to produce Deere's General Purpose tractors. When the company concluded that tricycle gear was the best way to proceed, it assembled twenty-three tricycle prototypes as GP-Wide Tread development models, and used the 5.75x6.00in horizontal two-cylinder engine.

1935 Deere Model B
The Model B was announced in late 1934, as the two-plow rated companion to the three-plow Model D or new Model A, introduced a year earlier. Delivered on steel, the Model B weighed 3,275lb while the Model A weighed slightly more than 4,000lb whereas in 1935, the Model D weighed 5,269lb. This tractor is owned by Herc Bouris.

that bad phone lines would confuse C or D, so the GP General Purpose was created. But testing pointed out problems.

In 1927, 104 prototypes were built. The C produced less power than IHC's new Farmall. And while Deere was still testing the C, International introduced an even more powerful engine. The wide radiator and front wheel track caused visibility problems. The resulting GP Wide Tread (GP-WT) more closely resembled the Farmall. Its rear wheels fit outside two rows and its narrow track front wheels fit between them. The C never went into regular production, but the GP did in 1928, and the GP-WT was introduced in 1929.

Enlarging cylinders increased horsepower in 1931. A tapered hood improved visibility in 1932 and allowed for "over the top" steering that reduced reduce front wheel steering whip.

Even as engineers solved problems plaguing the GP, they began two new machines. The Model A tractor (16/23hp) was introduced in 1934; the B (9/14 hp) came out in 1935. The A was a two-plow tractor with adjustable tread width; the implement hitch and PTO shaft were placed on the tractor's centerline; a hydraulic power lift system replaced the earlier mechanical lift. And the two-cylinder engine was re-engineered to burn almost any fuel: starting on gas, it was switched over when operating temperatures were reached.

1935 Deere Model B
Pneumatic rubber tires were introduced for the farm in 1931 and were optional—for between $50 and $200 per set—from most manufacturers within a year or so. That cost replaced the longer-spoke steel wheels with shorter-spoke rims designed to grip the tire. Some farmers simply cut the original wheels and welded on a new rim. While it saved them money, it proved to be a problem during World War II when rubber was unavailable.

1935 Deere Model B
All John Deere tractors at that time still ran on kerosene, sometimes known as "stove top.". It cost the farmer about six cents per gallon and made operating a John Deere tractor very economical. The Model B used the same four-speed transmission offered with the Model A, yielding a top speed of 6.25mph.

The B was described as two-thirds of the A. For 1937, high-clearance models were offered as well as a regular configuration, standard tread, and special versions for industrial use and for orchards. These featured fully enclosed rear wheels, independent rear wheel brakes for tighter maneuvering among trees, and air intake and exhaust stacks flush mounted on the tank cover. But in 1937 design changed significantly.

Deere's D, A, and B tractors were utilitarian. Experiments had produced striking looking body work on orchard models, frequently referred to as "streamlined." Deere engineers wanted all the machines more aesthetically pleasing. Henry Dreyfuss and other independent designers specialized in good design. For Dreyfuss, this meant utility and safety of the product, ease of maintenance, cost to produce, sales appeal, and product appearance.

Dreyfuss' group examined the A and B tractors and in November 1937, his version was unveiled. They enclosed the steering shaft, incorporated a grille and radiator cowl, and narrowed the width for better visibility. Deere advertised them as "Tomorrow's tractors today!" Styling consciousness spread, and tractors that date before 1937 are generally referred to as "unstyled."

1935 Deere Model B
Tests on the Model B were completed at the University of Nebraska in April 1935. The horizontal twin-cylinder engine displaced 4.25x5.25in bore and stroke. At 1150rpm, it produced 11.8 drawbar horsepower and 15.1hp on the pulley. Rear tread width was adjustable from 48in to 84in. Regular, industrial, and orchard versions were offered.

Another variation of the B-Orchard specifically met West Coast needs. Some 1,675 Model BOs were converted to crawlers by Lindeman Manufacturing Company of Yakima, Washington. (Deere bought Lindeman in 1945 and in the early 1950s moved all crawler production to its new plant in Dubuque.)

More work was possible with machines than with horses. As farmers acquired more land, Deere introduced more powerful row-crop machines. The 1938 Model G was a three-plow tractor. Where the B had been a scaled-down A, the G was scaled-up. The 1942 version, "modernized," included a six-speed transmission and was called the GM and could run on any fuel.

A small prototype tractor, the Model Y, was completed in 1936. Meant for the last small farms that still managed

1938 Deere Model A and 1949 Deere Model A
Bruce Henderson's 1938 Unstyled Model A hides many of the distinctive "improvements" in Arlo Schoenfeld's Styled 1949 Model A. New York industrial designer Henry Dreyfuss was hired to "make John Deere tractors more salable." By the time Dreyfuss and his colleagues finished with the Model A and B, they had improved everything from operator's seating to the appearance of the grille.

1941 Deere Model BO Lindeman
Jesse Lindeman produced the steel-tracked crawlers on Deere & Company's Model B-Orchard chassis for almost eight years. Near the end of the run, Lindeman was asked to help develop the successor, the Model M crawler. By that time, Deere had bought Lindeman's company and soon afterwards, moved all crawler production to its new plant in Dubuque.

with horse teams, it used an 8hp air-cooled Novo engine. Another twenty-three were built. The Y was uprated and renamed the 62 and another 100 were completed. Once perfected and renamed the Model L, Deere sold about 4,000.

The L engine and driving position were offset from each other, better to see the work to be done and the row to be driven. It used a Ford three-speed transmission and a foot clutch. Its two-cylinder engine was mounted upright. Deere marketed it to some buyers as their first tractor and to others as their second.

1941 Deere Model BO Lindeman
Lindeman used the standard Model B chassis and engine. The small horizontal two-cylinder engine was tested at Nebraska in a row-crop version. The 4.50x5.50in bore and stroke engine produced 17.6 drawbar horsepower and 19.7hp on the belt at 1150rpm. The styled B on rubber tested at 3,900lb; the crawler on 10in steel tracks weighed 4,420lb.

1941 Deere Model BO Lindeman
While Jesse Lindeman meant them as steps, they are so hard enough to find these days that owner/restorer Mike McGarrity of Pinion Hills, California, has labelled the Lindeman steps as "No Step." Lindeman manufactured crawlers from 1939 until 1947, producing in the end something like 1,675 of the crawlers.

By 1939, Henry Dreyfuss did a stylish grille and a striking curved cowl that became part of Deere design history. Then came the 1939 Model H as the last nail in the farm horse's coffin. Styled narrow-track and high-clearance versions were subsequently introduced.

Washington rationed steel and rubber during World War II. Still, development and experimentation continued so new products would be ready in peacetime. A new plant was ready, as well; production began at Dubuque on replacements for the H and L models. The 1947 M introduced Touch-O-Matic hydraulics, the Quick-Tatch mounting system, and reintroduced the vertical engine mount. The seat was adjustable, air-cushioned, and fitted with a padded seat back. Henry Dreyfuss created a steering wheel that telescoped a foot for easier driving when the farmer stood. The MC crawler, the Lindeman replacement, arrived

1941 Deere Model BO Lindeman
A variation on the Model B theme was executed by Jesse Lindeman and his brothers Joe and Ross in Yakima, Washington. Lindeman brothers experimented with crawler tracks on Deere's Model D and GP. But to Jesse Lindeman, it looked as if the Model B-Orchard had been designed to accept his crawler tracks.

1948 Deere Model D Styled
Weighing in at around 7,050lb without ballast for its University of Nebraska Tests, the styled Model D operated on Deere's horizontal two-cylinder 6.75x7.00in engine. At 900rpm, the engine produced 34.5 drawbar horsepower and 40.2hp on the belt pulley. Kenny Duttenhoeffer of El Cajon, California, owns this powerhouse.

in 1949. Electric start was standard on all the Ms, electric lights were optional.

Deere was near the limits of reliable power from two-cylinder slow-speed kerosene burners. Prototypes with

1948 Deere Model D Styled
Henry Dreyfuss and Associates didn't get to the venerable Model D until the 1939 model year. While the rest of the Deere tractors adopted a "streamlined" grille treatment with a strong vertical element with horizontal slashes, Deere's workhorse remained clean and powerful looking. Electric starting was optional.

1948 Deere Model D Styled
The final Waterloo Boys weighed nearly 6,200lb and measured 132in in overall length. Deere & Company's Styled Model D was 130in long, only 6in narrower (at 66.5in) and not quite 2in lower (at 61.25in) than the Waterloo. And weight? With electrics and hydraulics, with wheel weights and ballast, the D weighed 8,125lb and could pull nearly three-fourths the weight of the Waterloo Boy—4,830lb in low gear.

diesel engines were built. Thousands of hours of tests gave decisive proof. Introduced as the R in 1949, Deere's diesel—like Caterpillar's—started with a gas two-cylinder engine. The gas exhaust heated the diesel cylinders, making even cold weather starting easier.

Independent rear brakes and the padded seat were continued from the M. But live PTO was optional as was a new hydraulic system, Powr-Trol, operated from the driver's seat.

Beginning in 1952 when the 50 and 60 Series replaced the B and As, new tractors fit progressive niches. Duplex carburetion—one barrel per cylinder—came along with "cyclonic fuel intake" to feed equally and mix optimally the fuel and air in each cylinder. Power steering using on-board hydraulics was optional in 1954.

Then in 1956, after five new tractors in three years, Deere replaced the entire line. Some differences were immediately visible—a stylish new green-and-yellow paint

1949 Deere Model B Styled
When there's corn to be harvested, everything works, even the restored antiques. Arlo Schoenfeld of Charter Oak, Iowa, left his 1949 Styled Model B to wait for the harvester working nearby. Styled Model B tractors remained in production until early June 1952.

1949 Deere Model B Styled
Deere & Company introduced the final production series of the Model B in 1947. The horizontal two-cylinder engine was the 4.69x5.50in version that produced 19hp on the drawbar and 24.5hp on the belt pulley at 1250rpm. An electric starter was standard. The tractor frame was pressed steel, increasing strength and decreasing weight.

1949 Deere Model B Styled
In tests at the University of Nebraska, the last series of Model Bs pulled a maximum of 3,353lb in the lowest of its six gears. It weighed only 4,058lb. In sixth gear—transport gear—the tractor was capable of 10mph. The box beneath the operator's seat contained the battery.

scheme; others were more subtle as horsepower increased about 20 percent throughout the line. Powr-Trol provided a new higher speed lift and drop for implements that could take advantage of quicker maneuvering. Load-and-depth control met Deere's interpretation of the Ferguson system for automatic draft compensation. A remarkable option used engine power to change rear track.

These began the end of an era. The two-cylinder "Poppin' Johnnies," produced since the Waterloo Boys, were to be replaced. In one of the best kept secrets of manufacturing, work began in 1953 on new engines.

Seven years later, on Tuesday, August 30, 1960, John Deere's four- and six-cylinder engines surprised Deere's 6,000 dealers. The "New Generation of Power" promised 36hp to 84hp.

The Board's fear of silencing the "pop" was unfounded. In the next decade, more than 400,000 of these tractors sold.

1952 Deere Model MC
Bob Pollock's 1952 MC Crawler sits in contradiction to modern thought regarding tillage. With Pollock's as-yet unrestored John Deere Model 4A plow, the MC crosses the low-till contour plowing that represents the more prudent soil conservationist practices. But on this angle, the crawler catches the late afternoon western Iowa sunlight better.

1952 Deere Model MC
The MC was offered from 1949 through 1952. Tested at Nebraska, the 4,293lb tractor was powered by the Model M 4.00x4.00in vertically mounted two-cylinder engine, coupled to the four-speed transmission. Maximum horsepower was achieved at 1650rpm, and measured 20.1hp on the belt and 18.3hp at the drawbar. Standard equipment was 10in steel-track shoes.

1957 Deere Model 720 Hi-Crop
Introduced in 1956, the Model 720 Hi-Crop offered 32in of ground clearance. Powered by Deere & Company's horizontal two-cylinder diesel, the 720 produced 40.4 drawbar and 56.7 belt horsepower at 1125rpm. Engine displacement was 6.125x6.375in bore and stroke. In transport gear, the 720 topped 11.5mph, a dizzying speed when the operator's head is more than 10ft above the ground.

1957 Deere Model 720 Hi-Crop
When the scale-model toy manufacturer Ertl wanted to reproduce a Model 720 Hi-Crop, they came to Bob Pollock of Denison, Iowa, and measured and photographed his life-size giant. Months later, a box arrived containing a gold-plated commemorative 10in tall model of Pollock's 101in tall sugar cane tractor.

Chapter 5

Ford

Harry Ferguson's contribution to farming, the three-point hitch, was simple, effective, and economical. But getting Ferguson's invention to the farmer was even more significant. That credit went to Henry Ford.

Henry Ford's farm tractor grew out of agricultural history. Heavy, complicated, and expensive steam tractors had to move their own weight as well as pull plows. Ford wanted his tractor to be light yet strong, simple to operate, and cheap to buy and maintain—cheaper even than horses. He succeeded: a government test with his Fordson tractor concluded that farmers spent $.95 per acre plowing with a Fordson; feeding eight horses for a year and paying two drivers cost $1.46 per acre.

Ford was born in 1863, the second generation Irish-American born onto farmland outside Dearborn. But he preferred tinkering with machines. By 1896, his first machine ran and in 1903 his motor company began. He began serious tractor experiments late in 1905, and worked until 1907 on a prototype using his Model B car engine. Two more prototypes were tested through the fall and winter. In July 1917, two years after leaving Ford Motor Company to pursue his tractor interests, he incorporated his new tractor company, Henry Ford & Son. (Minnesota entrepreneurs, employing a man named Ford, named their product after the employee. This firm had tied up the Ford tractor trademark, to capitalize on the confusion.)

1918 Fordson (Left) and 1919 Fordson (Above)
Henry Ford began experiments with tractors before 1907 but he wasn't satisfied with the results until 1916 when, incorporated as Henry Ford & Son, he introduced his Fordson tractor. Mass produced on an assembly-line similar to his cars, Ford's tractor redefined the American farm tractor. This 1918 example, left, is owned by Daniel Zilm of Claremont, Minnesota. Above, owner Fred Heidrick, right, and friend Buren Craling pause for a moment after covering Heidrick's 1919 Fordson. The tractors were always sold new with the canvas cover, to protect the wood steering wheel and the wood induction coils. Few of them survived even well enough to make a pattern to reproduce.

Ford sampled his competitors' tractors, testing them on the family farm. He took them to his new Dearborn plant for the staff to examine. Because efficient manufacture was as important to Ford as product affordability, his tractors were designed to be strong enough to support the entire machine without needing a separate frame. Each of these units was run on rails to a central point in the factory for final assembly.

Newspaper and magazine stories reported the first fifty tractors produced were at Ford's farm for testing. The world learned that Ford's dream was approaching reality.

War broke out in Europe. Germans sunk a ship a day and England rapidly lost food, able-bodied farmers, and draft animals. The British War Mission came to see the tractors. Very much impressed, they returned to encourage farm tractor production back home.

1919 Fordson
Henry Ford's first experiments with farm tractors were based on his automobiles—in fact he called his first one in 1903 an Automobile Plow. The pressures of the automobile business kept him from much more than experimentation until his Model B plow in 1915. The Fordson tractor entered production in December 1917—and was guaranteed success with a sale of 7,000 to the United Kingdom.

On April 6, 1916, the United States entered the war. The next day, Britain telegraphed Ford to borrow the engineering drawings for the Government "in the national interest." Ford agreed and by mid-May parts, patterns, implements, and six engineers were sent to find a suitable factory.

But London was bombed; all proposed tractor plants were rushed into war-plane manufacture. Tractors had to come from America.

Britain first ordered 5,000 Fordsons at cost-plus-$50, about $700 each. First delivery was to be within sixty days! Limits in available shipping space slowed delivery but a total of 7,000 were there by spring of 1917.

British agriculture was ready. In Ireland, Harry Ferguson, an aviator and auto racer, had taken the cause of the tractor to his heart, selling the Overtime, the British Waterloo Boy.

Ferguson, fourth child of James Ferguson, was born in 1884 near Belfast. Steam traction engines fascinated him and anything mechanical easily lured him from the farm.

1919 Fordson
The Fordson remained in production in the United States until 1928 when a price war began against International Harvester and all the other makers decimated the competition and finally did in Ford himself. But a market continued for the tractor in the United Kingdom and the Fordson—albeit a modified version—remained in production in Cork, Ireland, until 1946.

As Europe's war approached the United Kingdom, the need for food met Ferguson's love for machines. The Irish government hoped to improve tractor performance and asked him to visit farmers for educational demonstrations.

What impressed Ferguson most was the inefficiency and danger in the single-point plow hitch. The risk of hanging up the plow on a hidden rock was great. Horses

1919 Fordson
The Fordson first appeared with the rear-drive worm gear above the rear-wheel axle. But this caused lubrication problems which overheated the gear and heated the housing. This heat transferred out the seat post and in summer, farmers complained the seat was just unbearable. Later versions reversed the gear position and bathed it in oil.

1919 Fordson
One of Ford's greatest accomplishments was in making a tractor without a traditional frame. The engine, gearbox, and front axle were each integral elements—making up the unit frame—and each was a load-bearing member. This provided the tractor great strength while eliminating extra weight. The Fordson weighed around 2,500lb.

simply stopped moving. Tractors' spinning engines and flywheels kept going in a motion that often wound the tractor around the final-drive gear, bringing the tractor nose up and over on itself.

Even if the impact did not flip the tractor, the plow was usually damaged after impact. Ferguson also noticed that while using implements originally designed for horse-drawn farming, tractor farmers rarely got satisfactory results. Over uneven soil, the draw bar rose and fell as the tractor moved along; the furrow height either needed constant adjustment or it simply ended up sloppy and uneven. Harry Ferguson understood tractor makers' wish for great weight. It not only kept traction to the drive wheels but it held down the plow.

Back in Dearborn, Henry Ford & Sons tractor company was busy. In April 1918, daily production was sixty-four. By that July, 131 Fordsons rolled out of the Dearborn plant every day.

Fordson shortcomings surfaced quickly. Early design set the final drive right below the farmer's seat. This inefficient worm-gear system generated as much heat as power and this heat transferred up the farmer's steel seat. Later versions inverted the system and bathed the worm in oil, cooling the system and the farmer's backside. By 1920, these problems had largely been solved and sales, always steady, reached 70,000 near year end.

1919 Fordson
Ford's four-cylinder upright L-head engine measured 4.00x5.00in bore and stroke and was rated at 10 drawbar and 20 belt pulley horsepower at 1000rpm. Spark was provided by a low-tension magneto in the flywheel that fed current into the "buzz box" induction coils, shown with the metal cover removed. The magneto produced between 6 and 14 volts depending on engine speed. Below, a type of inclinometer, known as the "mag cut-out," shut off spark when the tractor nose lifted.

Ford moved Ford & Son tractors into his car company's new River Rouge plant. Production reached a record 399 a day, 10,248 in September. But the war ended. Over-capacity caught up. Sales in the 1920–1921 depression dropped to 36,793.

To keep production up, Ford cut the price of his tractor again and again. He sought a price level that would put the tractor in everyone's hands and a production level to keep his factories busy. He took losses to meet those goals. His price war enraged and broke many competitors.

In Belfast, Harry Ferguson worked on his new wheelless plow. Then in December 1917, he learned that Ford planned a tractor plant in Cork, Ireland, and until production was running, several thousand more Fordsons were to be imported. Ferguson took drawings and raced to London

1923 Fordson Snow-Motors Conversion
Ford tractor collectors marvel at the variety of options available to Ford operators at the time the Fordson and even subsequent N Series tractors were in production. Fordsons powered mine trains, scoop shovels, and even a Snow-Motor or two. The exact production is unknown, and even the year of manufacture is an unsolved mystery.

to Ford's representatives. He explained his theory that efficient farm mechanization required the implement to become part of the tractor when hitched on, but readily detachable again. This concept changed tractor farming forever.

The result was the Duplex Hitch. Attached to the tractor by two sets of struts, one above the other, the plow resisted the tendency to rise after impact because the upper arms forced either the plow or the tractor nose down

1923 Fordson Snow-Motors Conversion
A wide-angle lens exaggerates the perspective of a snowflake's-eye view of the drive mechanism of the Fordson Snow-Motor. The chains spun the spirals that drove the Fordson. The steering wheel was connected to the differential and the Snow-Motor steered more like a Cletrac crawler.

1923 Fordson Snow-Motors Conversion
Final-drive chain gears emerge from the rear of the Snow-Motor's modified differential case. Spring-tension idlers keep the drive chains taut even though the frames supporting the spirals appear extremely solid. At the top, the steering gear swings forward or back to engage or disengage the drives.

harder. This meant that the plow itself could be lightweight. But as the tractor pivoted over changes in field surface, furrow depth changed opposite to what the front wheels did. There was no draft control. Not yet.

Ford saw Ferguson only as an innovative machinery salesman. Ferguson wanted Ford to start a plow plant in Ireland. Both men knew the plow needed some engineering clean-up. Without a depth-control device its usefulness was in doubt. Ford dismissed the whole thing.

The Fordson was a continuing success. Production rose to nearly 69,000 in 1922 despite a continuing postwar depression, and to nearly 102,000 in 1923. Healthy export business continued, and between 1920 and 1926 nearly 25,000 were delivered to Russia. Ford had entered the tractor market in 1917 and by 1921 he had hold of two-thirds of the entire market. But he had not only overwhelmed his competition, he had also educated them. From Ford, they learned about automobile-style mass production.

Ford, selling largely through his automobile dealers, had never recognized the value of his own implements. When Fordsons failed to perform as expected, the tractor was blamed, not the unmatched product mix. Despite reduced prices, production declined and by 1928, Ford quit selling Fordsons in the United States. International Harvester resumed the lead.

Ferguson continued, adopting an internal on-board hydraulic-lift system to aid row-end maneuverability. By adapting sensors to the hydraulics, replacing rigid mounts with ball joints, and increasing the top single-strut angle, his new three-point hitch with automatic draft control was ready.

But Ford was converting his Dearborn plant to Model A production. And in October 1929, the stock market crashed and the Great Depression began.

Ferguson, frustrated by Ford's withdrawal and the economy's failure, arranged with an English tractor maker for li-

1923 Fordson Snow-Motors Conversion
Educated guesses estimate this Snow-Motor Fordson's year of manufacture as 1923, since the latest patent date on the brass plate is a barely legible August 1, 1923. Its serial number—100-A—raises more questions than it answers.

1923 Fordson Snow-Motors Conversion
Out of its element in a field of early growth summer wheat, Fred Heidrick's unusual Snow-Motors Fordson conversion sits on blocks. Heidrick's farm does not get snow but he toyed with using the snowmobile to get around in the mud created by the wettest winter in more than a decade.

censing agreements and obtained a tractor for development. In late 1938, Ferguson and a small staff took his plow-tractor combination to Michigan. Ford had ceased US tractor production reluctantly ten years earlier. Fordson production continued first in Cork, then after 1933 in Dagenham, England. Yet Ford was thinking of a new, domestic Ford.

Ferguson's timing was perfect. Ford was disappointed with his own engineers' work and already knew Ferguson. Ferguson's tractor finished its demonstration. To better explain his three-point hitch and draft control system, Ferguson had brought along a scale model.

The two talked and after a while, shook hands. A partnership was formed. Conditions favored both sides, providing Ferguson a manufacturer for his idea and Ford an idea to manufacture.

Ford invested $12,000,000 in tooling costs and helped Ferguson finance his new distribution company. The 9N, known as the Ford Tractor with Ferguson System, was introduced June 29, 1939. The $585 price included rubber tires, power takeoff, Ferguson hydraulics, an electric starter, generator, and battery; lights were optional. The 9N sales brochures showed possible mounting points for a radio—due to the quietness of the engine!

Ford's 9N further improved the cantankerous Fordson by updating the ignition with a distributor and coil. An innovative system of tire mounts for the rear wheels and versatile axle mounts for the fronts enabled farmers to accommodate any width row-crop work they needed. Ford had aimed for the perfect tractor with his 9N. He had tried before with the Fordson and believed he had it right by the time he introduced this new tractor.

With the introduction late in 1942 of the 2N, Ford incorporated farmers' suggestions for improving the tractor. Other changes occurred because of wartime needs for metal and rubber. Yet the new war offered a windfall marketing

Previous page
1923 Fordson Snow-Motors Conversion
Set by the roadside only for photographs, this Snow-Motors Fordson conversion was used in the 1920s and 1930s by James McIvor, who was a US Mail contractor. He used the Fordson and its sled wagon (runners are hidden in the tall grass) to deliver the mail and supplies when winter snows closed the roads between Truckee, California, and North Tahoe, Nevada.

1923 Shaw Ford Model T Conversion
Shaw Manufacturing Company of Galesburg, Kansas, produced kits to convert Ford's Model T or A automobiles into a tractor. Most conversions failed because the automobile radiators were not large enough to cool the engine under constant load. Shaw's kit included front and rear wheels and other bits. Owner Fred Heidrick painted Shaw's parts red in contrast to the original Ford.

opportunity for Ford. The US Government Office of Production Management proposed cutting tractor production 20 percent and increasing repair and maintenance parts by 50 percent. Ford claimed his lightweight tractor used one ton less steel than existing large tractors. He suggested that nearly 500,000 older machines could be scrapped and replaced with his 9N. Half a million tons of steel would be enough for several battleships. Ford would make all the tractors for America.

1941 Ford Model 9N
Keith McClung of San Juan Capistrano, California, runs a cloth over his 1941 Ford 9N before putting it to work. The 9N remained in production until its successor, the 2N, was introduced in 1942. Almost all 9Ns were sold on pneumatic rubber although World War II forced most 2N models on to steel.

His competitors wailed. The OPM declined.

The 9N and 2N certified the engineering ideas Harry Ferguson had struggled to prove for decades. Yet when Ford introduced the 8N, Ferguson's name was no longer on the tractor.

Henry's grandson, Henry II, 26 years old, was called home from the Navy to run the company. He soon learned that tractor operations had already lost $20,000,000. Henry II set out to cut the costs.

Henry Ford had hoped to clarify his agreement with Ferguson. But the frustration continued. The prospects of reconciliation seemed slim. In 1946, Ford Motor Company tried to buy into or buy out Ferguson to form a new sales and distribution company. The terms offered less and less to Ferguson: 30 percent of the new company and no royalties on the Ferguson system patents.

1941 Ford Model 9N
Ford quit the US tractor market in 1928 after a brutal fight. When he returned in 1939, his tractors wore a new badge up front and an ingenious apparatus at the rear. Irishman Harry Ferguson agreed on a handshake to market his implement hitching system with Ford's new tractor, called the Model 9N.

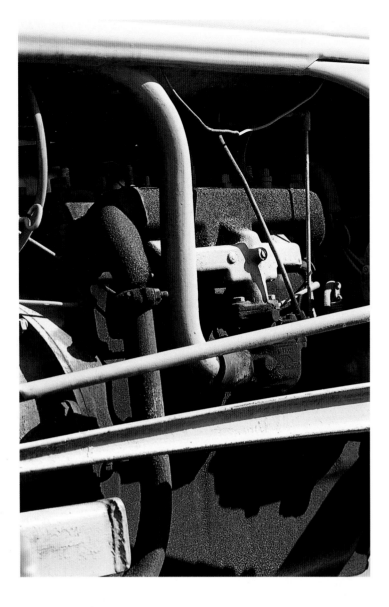

1941 Ford Model 9N
The 9N used Ford's vertical four-cylinder L-head engine with 3.18x3.75in bore and stroke. In its Nebraska Tests in April 1940, it produced 12.6 drawbar horsepower at 1400rpm. and 23.1 belt pulley horsepower at 2000rpm. It was sold with electric starter, generator, and headlights as standard equipment.

1941 Ford Model 9N
The 9N weighed 3,375lb and was equipped with Ford's three-speed transmission. Top speed was just about 7.5mph. It was sold on 4.00x19 front tires and 8.00x32 rears. The 9N also introduced on-board hydraulics, operated by the lever below the right side of the operator's seat.

Their Handshake Agreement was dissolved. Effective December 31, 1946, Ford would continue to manufacture tractors through June 1947 for Ferguson to market, but Ford immediately established its own distribution company, the Dearborn Motor Corporation.

Then Henry Ford died on Monday, April 7, 1947 at age 83. In his lifetime, he had sent 1.7 million tractors out of his factory doors with his name on them.

In July 1947, right after Ford's last shipments to Ferguson, the new 8N was introduced. It boasted some twenty improvements over the 2N, including a four-speed transmission. It came equipped with the Ferguson System. No royalties were paid.

Ferguson filed suit against Ford for $251,000,000, charging "conspiracy" to infringe on patents and other complaints. Ford denied or repudiated every allegation. The legal battle, not merely a patent suit but now an antitrust suit, dragged on for four years. Ford counter-sued in July 1949, charging Ferguson with "conspiracy" to dominate the world tractor market and other complaints. The trial dragged on for months. In the end, Ferguson received a $9.25 million settlement from Ford on April 9, 1952.

Ferguson's patents on much of the three-point hitch had run out by the time the suit was settled. But Ford had already developed its next tractor to commemorate its 50th Anniversary in 1953. The new NAA tractor began production shortly after the New Year. Substantially restyled, it was officially named the Golden Jubilee 1903–1953 Model, but quickly became known as the Jubilee. It was 4in longer and higher, and about 100lb heavier than the 8N. It also introduced Ford's new Red Tiger engine, an overhead-valve four. Live Power Take Off was optional. The Jubilee, a three-plow tractor, was produced until 1955, when Ford changed its tractor direction.

1947 Ford Model 2N Cotton/Cane Special
This is the operator's-eye view of the 1947 Ford 2N Sugar Cane/Cotton Special. A single front tire was fitted instead of Ford's standard wide front end. Just visible past the wide rear fenders are 9x40 rear tires on special rims which allowed 4in of adjustment.

1952 Ford Model 8N With Funk Brothers V-8 Conversion
There was no Nebraska Test. Palmer Fossum's "hot rod" began life as a normal 1952 Model 8N, but an earlier owner wanted three-plow power from a two-plow tractor. Funk Brothers Aircraft in Coffeyville, Kansas, offered a 100hp flathead V-8 conversion. Farmers who knew the pop of a two-cylinder or the rumble of a four- or six-cylinder exhaust were startled by the bark of a V-8 with twin pipes.

1958 Ford Model 501 Workmaster Offset
It is no optical illusion nor is it a wide-angle lens distortion. In fact, the tractor is called the Model 501 Workmaster Offset. The engine was mounted left of the tractor centerline to improve operator visibility working with cultivators mounted beneath the tractor.

Until the introduction of the new 600 and 800 Series—five tractors in all in 1955—Ford had been a one-tractor company since 1917. Now Ford, which had on and off again held the major market share, could compete more effectively against all its opponents.

Liquefied petroleum gas (LPG), became a fuel option in the United States, and in the United Kingdom diesel engines were available for the Fordson Major. Workmaster and Powermaster tractors were introduced in 1958, when diesel engines also came to the United States. In 1959, Ford introduced its "Select-o-Speed" transmission using hydraulic power for gear change in an automatic-type transmission.

Harry Ferguson died in his home on October 26, 1960. In late summer, he talked about getting back into the game. He wanted to build a new tractor, a tractor that would make use of the torque converter automatic transmission and four-wheel-drive.

1958 Ford Model 501 Workmaster Offset

1958 Ford Model 501 Workmaster Offset
The Workmaster series included a 501, 601, and 701. The Model 501 used Ford's four-cylinder 3.44x3.60in bore and stroke engine. The Offset offered nearly 30in of ground clearance, almost enough to clear the tall grass of one of Palmer Fossum's Northfield, Minnesota meadows.

Chapter 6

International Harvester

International Harvester's tractor division was born like most others—after decades of producing other implements, the tractor became a necessary addition to the catalog.

In July 1831, Cyrus Hall McCormick, born of Scottish-Irish stock, was an inventive 22 year old when he demonstrated his first reaper. Cyrus' father, Robert, was a dreamer and tireless inventor. For the first seven years of Cyrus' life, Robert tried to build a mechanical reaper. Ignorant of failures elsewhere in the world, Cyrus went on to achieve his father's dream.

Cyrus slowly "invented" a reaper, modified it, corrected it, and in 1834, patented it. He traveled to promote it and to find others to make it. When shipping each $100 reaper to the West meant an additional $25 in freight, McCormick decided to move himself instead. After selling 123 machines in 1846, he moved to Chicago in 1847.

Gold in California moved thousands of farmers from the East and Midwest. Markets for McCormick's reapers increased. His next decade was a challenge, beginning with the lantern kicked over by Mrs. Patrick O'Leary's cow on Chicago's North side. When the embers cooled, the Chicago fire of 1871 had destroyed more than $180 million in property over 3 1/2 square miles, taking with it McCormick's factories and offices. McCormick was 62 and tremendously wealthy. But instead of retiring, he rebuilt.

1954 International McCormick Farmall Model 300
The McCormick Farmall 300 was introduced in 1954, and brought to the farmer International's new Fast Hitch system with full on-board hydraulics and the Torque Amplifier producing basically two speeds in each gear. The 300 Series was produced until 1956. Optional rear tires were easily fitted, these being such an example: 12.4x38, to increase the tire-patch on the ground. With standard tires, the 300 pulled a maximum weight in low gear of 4,852lb in its tests at the University of Nebraska in May 1955. The type C169 engine produced an exceptional 283lb-ft of torque at 1750rpm. This high-clearance example is owned by Bob Stroman of Hawthorne, California.

The pace of the next decade proved too much and McCormick died in 1884. He was succeeded by his son Cyrus, Jr. Within months, McCormick's legacy turned into organized labor confrontations that led to riots. But young Cyrus had already spent five years as his father's secretary and at 25, he took over its worldwide operations.

A former dry-goods manufacturer and merchant posed McCormick Sr's last challenge. This other New England refugee who arrived in Chicago in 1873 was William Deering. Through the next six years, McCormicks battled Deering and other competitors for harvester and binder markets. Mergers were discussed. An attempt surfaced in 1897 when the Deerings offered their company to young McCormick. But McCormick was too far extended keeping ahead of the Deerings while developing European markets. For Deering the

Previous page
1917 International Mogul Model 8-16
International introduced the Model 8-16 Mogul in 1914 but fewer than two dozen were assembled that first year. The tractor remained in production through most of 1917, by which time slightly more than 14,000 had been produced. Its very narrow front axle and its high frame offered Mogul operators much greater maneuverability than with other makes.

1917 International Mogul Model 8-16
The air intake, capped in silver, stands far above the dust kicked up by the forward-facing exhaust below the engine.

problems were similar; expanding production to challenge McCormick had strained their finances as well.

By 1902, Deering was much more self-sufficient, its manufacturing more efficient. Yet the competitive battles during the past decade had led to overproduction and overpopulation. Hustling for every sale, its stock on hand far exceeded realistic needs. Whenever one competitor opened a sales branch, others opened in the same locale. By 1902, there were more than 40,000 dealers.

On August 12, 1902, $60 million changed hands. The International Harvester Company acquired the factories, warehouses, inventories, and properties of Deering Har-

1917 International Mogul Model 8-16
Adjustments were possible to the fuel mixture, based on whether or not the engine was warm or cold. The Mogul used a single-cylinder horizontal engine coupled to a one-speed forward or reverse transmission. Top speed was 2mph in either direction.

vester, McCormick Harvesting Machinery Company, and others. By 1906, in Upper Sandusky, Ohio, the new conglomerate produced its first tractors.

International Harvester's first tractors used 15hp horizontal one-cylinder engines mounted on rollers on the frame and shifted forward or backward to engage friction drive. In 1906, IHC produced twenty-five of these for testing and development. In 1907, 200 more were built. Friction drive eventually proved unsuitable and when tractor production was transferred to the former Aultman-Miller works in Akron, geared transmissions were used.

International Harvester, like companies before and after, attempted to maintain separate lines of production for each of its former independents. Between 1909 and 1914,

1917 International Mogul Model 8-16
The high-cut front frame allowed full clearance for the narrow track front wheels to turn more than 45 degrees in either direction. Steering was extremely simple: from the steering wheel by a straight shaft to a horizontally mounted worm-and-sector arrangement. Despite its tight turning ability, cranking the wheel around so many turns still made its great maneuverability into hard work.

1917 International Mogul Model 8-16
Viewed between the seat post on the left and the right rear wheel, the externally mounted rocker arms, tappets, valve stems, and pushrods were subject to all the dirt they could suffer. Ignition spark was produced by an oscillating magneto. Bore and stroke was 8.00x12.00in and rated power was reached at 400rpm.

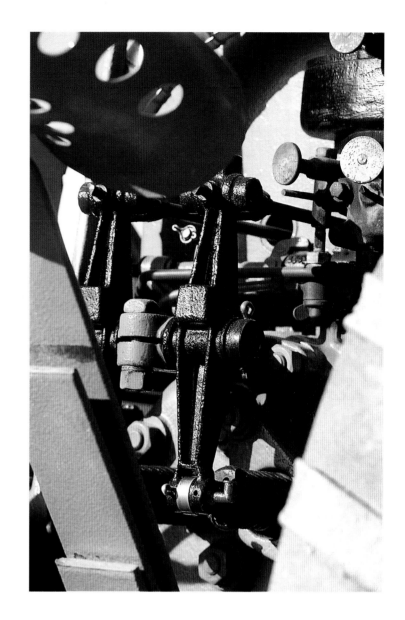

IHC produced the heavy 20hp Mogul for sale by McCormick dealers. For Deering distributors, it introduced the Titan, rated at 27-45hp in 1910. International's goal of equal products for separate divisions was tripped up. The designers of the Mogul benefited from their experiences before undertaking the Titan.

The Winnipeg trials were inaugurated in 1908. The Canadian market was large and influential and a good performance there guaranteed sales. Horsepower escalated until in the 1912 exhibitions, IHC linked three Titans together, pulled fifty-five plows and turned a swath 64ft wide. Meanwhile US farm magazine and newspaper writers called repeatedly for smaller machines. The giants were too expensive and too hard to maintain. The average farmer had little mechanical experience.

1918 International Model 8-16 Kerosene
IHC's International 8-16—with its fully enclosed engine—was a radical departure from the appearance of Titan and Moguls that came before. The 8-16 was introduced in 1917 and remained in production until 1922. This 1918 example is owned by Joan Hollenitsch of Garden Grove, California. It sees occasional use at the Antique Gas and Steam Engine Museum of Vista, California.

1918 International Model 8-16 Kerosene
International used its own inline vertical four-cylinder 4.25x5.00in engine. Tested at Nebraska in 1920, the engine produced 11 drawbar and 18.5 pulley horsepower at 1000rpm. To improve forward visibility, International placed the radiator at the rear of the engine. This naturally improved operator warmth for winter weather operation. The rest of the year, however, must have been uncomfortable.

1920 International Titan 10-20
When the Titan was introduced, International Harvester fitted shorter rear fenders that stopped just about at the operator's seat. Beginning in 1920, full-length fenders were used. In 1920, the Titans sold for $1,000. The red wheels became a trademark of International Harvester tractors.

1920 International Titan 10-20
IHC introduced the 10-20 Titan in 1915, producing only a handful that year. Between 1916 and 1922, the company manufactured more than 78,000 of the strong, straightforward machines. Cooling without a radiator and fan was managed by thermosyphon, which percolated the hot water from the engine into the high-mounted, gray water tank.

In 1914, IHC's kerosene-fired single-cylinder 8-16 Mogul was introduced. This was followed in 1915 by the 10-20 twin-cylinder Titan. The economic timing was good. The threat of war in Europe created an additional market for food and fiber.

As tractor markets settled, IHC reorganized following an antitrust suit that charged IHC with maintaining dual dealerships in areas where both McCormick and Deering outlets still existed. Eight thousand dealers were let go by 1919. Products were combined, as McCormick-Deering. The remaining 13,800 dealers broadened their bases, each carrying McCormick and Deering products. IHC was left leaner and more streamlined to enter the 1920s.

That American farmers were ready for leaner, smaller tractors was undeniable. Tractors in 1910 averaged 504lb

1920 International Titan 10-20
Owner Bill Cue cranks over his 1920 Titan. His tractor is one of 21,503 Titans manufactured in 1920, the peak production year.

1920 International Titan 10-20
The Titan was the 23rd tractor tested when the University of Nebraska began evaluating farm tractors. The Titan used a horizontal two-cylinder engine in which both pistons moved in parallel—that is, both sparked at the same time. Vibration was dampened by a huge heavy flywheel. Oil flow could be confirmed visually.

1920 International Titan 10-20
Engine displacement was 6.50x8.00in It produced 9.9 drawbar and 28.2 pulley horsepower at 575rpm. The tractor weighed 5,708lb when tested at Nebraska. Its maximum drawbar pull was 2,660lb in low gear at slightly less than 2mph. Top speed was 2.9mph.

per horsepower. IHC took the tractor production lead, ahead of Rumely and Hart-Parr. Competition was still fierce. By 1917, Henry Ford shipped 7,000 small Fordsons to England. IHC's 1918 International 8-16 weighed 3,300lb, or 206lb per hp. Ford's 1918 Fordson weighed 2,700lb, or 150lb per hp.

In 1918, with war in Europe and 7,000 sales in the UK, Ford attacked the home front. Fordson passed IHC's tractors in production. In 1920, when totals from all makers hit 203,000, Ford produced nearly three-quarters of that number.

To force a showdown, Ford cut his price by $230, to $395. IHC was forced to follow. International Harvester fought back, cutting the price of the Titan 10-20 to $700 and the International 8-16 to $670, including a plow. It quickly consumed the tractor stock piles. Still, with the Fordson at $395, the advantage lay with Ford. And worse,

the price wars meant both companies sold below cost. (Some outside suppliers charged more than $395 just for their engines.) For IHC, without the auto revenues that Ford enjoyed, production costs had to be trimmed in ways never before imagined.

IHC's salesmen turned every Fordson demonstration into a field contest, and International's tractors won each time. By the end of 1924, IHC's sales increased.

When IHC replaced the International 8-16 in 1922, more than 33,000 had been sold. Introduced at $1,150, it was selling for less than half, including the two-bottom plow. Its replacement was IHC's first unit-frame tractor. Rear PTO was optional on the new 15-30 Gear Drive Tractor. (More than 99,800 were produced by 1929, when "The New 15-30" was introduced.)

In 1923, as IHC fought Ford, its engineers tested general-purpose tractors, targeted at the Fordson's failures: low

1935 International McCormick-Deering Farmall F20
McCormick-Deering Division of International Harvester Corporation began production of the F20 Farmall in 1932, and introduced the tractors on steel wheels. Rubber was offered by the end of 1934, but it was an expensive option, nearly $1,000 in addition to the $895 cost of the tractor with narrow front end.

ground clearance limited its use in corn and cotton crops to planting, and its wheel placement required too much maneuvering room.

The goal was a machine that would replace all the animal power on the farm. Engineering had tried a cultivator, configured as a reversed tricycle; its driving wheel also did the steering at the rear. But the cultivator wouldn't draw a sod-breaker plow through unbroken prairie. So they reversed the whole affair and moved the engine to near the middle of the channel-iron frame.

Another twenty of these "Farmall" prototypes were ordered, with every conceivable implement. Testing continued. Millions of dollars were spent. The first pre-production version was sold in Iowa in 1924. Two hundred of these sold for $825 apiece.

Dealer follow-up brought several successful ideas from farmers' own adaptations of the new IHC machine. The first Farmall had no model designation, and IHC executives resisted the temptation to label it with horsepower ratings. In 1925, the price was raised to $950; fenders were an extra $15. Production increased like untended weeds: at the end of January 1930, the Tractor Works turned out 200 machines a day and from February 20, 1924, when production was authorized, until April 12, 1930, 100,000 Farmalls were built.

In 1930, the Farmall adopted a name, the "Regular," when its new F-30 version was introduced and in 1932, IHC introduced a companion F-20 model, available on rubber, using the Regular's engine. From the start of the Farmall series, a full range of compatible implements was

1935 International McCormick-Deering Farmall F20
International used a four-speed transmission on its F20, with a top speed of 3.75mph. In its tests at the University of Nebraska, in low gear the rubber-tired gasoline engine version towed 2,927lb, comparing well against the 2,334lb towed by the gasoline-engine steel-wheel version tested in July 1934.

available. A rapid mounting system, called Quick Attach, meant that changeover took minutes. Everything from golf course fairway lawn mower gangs to plows or cultivators was offered.

The W-40, International Harvester's big tractor, arrived in 1934, with a diesel version. This WD-40 was the first US-built wheel-tractor to use Rudolph Diesel's technology. It started as a gasoline engine, hand cranked, and then the engine automatically switched over to diesel fuel.

As early as 1926, IHC experimented with crawler versions of the McCormick-Deering 15-30. Work progressed based on the 10-20, and a crawler—the TracTracTor—entered regular production in 1929. Fitted with steering clutches, it was basically a wheel-type chassis adapted for tracks. The T-20 TracTracTor became the working models for IHC crawlers. Foot brakes supplemented the steering clutches; square-corner turns were possible.

International Harvester asked Raymond Loewy to bring "styling" to its tractors in the 1930s, to clean up design and appearances of the machinery, dealerships, and even the corporate logo. When the new crawlers were introduced in late 1938, they benefited from Loewy's scrutiny and imagination. He raised exhaust pipes to clear the operators' heads, moved pedals and levers for easier reach and styled a

1935 International McCormick-Deering Farmall F20
The Farmall F20 was an extremely simple and extremely rugged machine. Weighing about 4,500lb, it was also strong and reliable. There are countless stories of F20s in use for fifty years, out of service only once to grind the valves.

distinctive radiator grille, which also carried over into the Farmall series.

The Farmall F Series was replaced over a two-year span by new A, B, H, and K models. The A's streamlined hood and offset operator's seat were Raymond Loewy's influence, and continued through the entire lineup. The H was supplemented by the M, a three-plow row-crop tractor. Both the H and M were also offered in high-crop versions, and a diesel engine was introduced for the M in 1941. A new small tractor was introduced in 1947: the Farmall Cub rated just 9 drawbar horsepower, good for a single 12in plow.

The mid-1950s brought complete reorganization to its tractor lines; changes and the diversity of models introduced by 1960 were abundant. Late in 1956, Farmall and

1935 International McCormick-Deering Farmall F20
When tested at Nebraska in November 1936, the Farmall was powered by International's own 3.75x5.00in bore and stroke four-cylinder engine. At 1200rpm. the gasoline engine produced 19.6 drawbar and 26.7 belt pulley horsepower. Its rust betrays the usefulness of this Farmall.

1941 International Model TD-14
International had begun producing crawlers with diesel engines by 1936. This 1941 Model TD-14 was powered by International's 4.75x6.50in four-cylinder engine. At 1350rpm it produced 51.4 drawbar and 61.6 brake horsepower (a similar rating to belt pulley horsepower where no pulley is fitted). In production for ten years, it weighed nearly 17,600lb.

Next page
1941 International McCormick Farmall Model H
Raymond Loewy's discipline accomplished for International Harvester what Henry Dreyfuss did for John Deere. Loewy cleaned up the tractor's lines, unifying its appearance. The Farmall H was introduced in 1939 as a row-crop tractor, and a panel was removable from the grille to attach a lever to "steer" Farmall's shifting gang cultivators.

International introduced a new color scheme, adding white to the grille and hood slashes. This series change was produced only until 1958 when another major restyling took place accompanied by model designation changes. In 1958, the grille was substantially redesigned, appearing more aggressive and forceful.

In the 1970s, the economy began to collapse when fuel prices rose and produce prices fell. By 1979, IHC lost the lead in tractor sales, this time to John Deere. Unrealistic federal farm policies kept grain prices up despite surpluses.

Tractor sales from all makers were only 29,247 in 1981, less than some single-year sales of any individual Farmall in the past. Dealers went under by the hundreds.

In 1982, IHC got serious about itself. It sold its construction equipment division, closed factories and consolidated operations. Then, in 1984, IHC surrendered its heritage. Tenneco, which already owned J. I. Case, bought IHC's farm tractor and implements divisions for $486 million. After that, Case-IH as it was now known, became Tenneco's largest division.

1941 International McCormick Farmall Model H
Late afternoon winter sunlight only slightly accentuates International Harvester's corporate red. Model H Farmalls shared wheelbase and other features with the Model M, although the H sold for $130 less—$962 on full rubber compared with $1,112—plus the electric starter and lights for another $50. Keith McClung of San Juan Capistrano, California, owns this tractor.

1941 International McCormick Farmall Model H
The Farmall H was offered on four steel wheels, four pneumatic rubber tires, or a mix. On rubber, its rear tires were 10x36 and its fronts were 5.50x16. Sold throughout World War II, rubber tires were discontinued until 1948. International fitted its own five-speed transmission to these tractors, offering a transport gear top speed of more than 16mph—a terrifying prospect on steel.

1941 International McCormick Farmall Model H
The Model H engine was International's 3.375x4.25in bore and stroke four-cylinder mounted vertically. By this time, IHC was producing its own magneto and carburetor as well. At 1650rpm, the engine produced 19.1 drawbar and 24.3 belt horsepower. Total weight of the tractor is about 5,550lb.

1941 International McCormick Farmall Model H
The Model H, shown here, weighed nearly 5,550lb, its companion Model M weighed 6,770lb. With gas engines, the M produced 26.2 drawbar horsepower and pulled 4,233lb, in low gear. The H produced 19.1hp and pulled a maximum of 3,603lb. The H returned 11.75hp hours per gallon of fuel where the M provided fuel economy of 12.16. The M was sold into 1952, the H into 1953.

Previous page
1946 International McCormick Farmall Model A
When the Farmall F20 went out of production in 1939, it was replaced by a new series of tractors that were more streamlined in appearance. The Farmall Model A offset the engine to the operator's left for improved vision.

1946 International McCormick Farmall Model A
The Farmall Model A sold for $575 in 1940. This model features optional lights and electric starting, which cost another $31. It remained in production until 1947 when it was replaced by the Super A, which differed most significantly in offering adjustable rear tread width.

1946 International McCormick Farmall Model A
The offset seating position was meant to greatly facilitate cultivation of smaller and more delicate crops. In fact, International named this offset position "Culti-Vision." The tractor is deceptive looking: despite its small appearance, it weighs 3,570lb.

1946 International McCormick Farmall Model A
Tested at Nebraska, the Model A "Culti-Vision" used International's four-cylinder 3.00x4.00in engine. Run at 1400rpm, it produced 13.1 drawbar and 16.3 belt pulley horsepower. In low gear, it pulled a maximum of 2,387lb. This sparkling example is owned by John Frazer of Escondido, California.

1946 International McCormick Farmall Model A
Making the best use of compact space, International fitted the belt pulley wheel and power takeoff (PTO) shaft beneath the operator's seat on the Farmall Model A "Culti-Vision." Originally, this was a $39 option but by the final year of production, it was standard equipment.

1954 International McCormick Farmall Model 300
The operator sits on the battery box, within easy reach of the Fast Hitch implement attachment and height adjustment controls. International Harvester—always a full-line equipment manufacturer—offered literally dozens of accessories and implements to use with the Fast Hitch.

1954 International McCormick Farmall Model 300
In standard trim, the Model 300 weighed 5,361lb and was fitted with 5.50x16 front tires and 11x38 rears. The 300 Series was offered not only in standard and row-crop fronts but also a Utility version and a high-clearance model. The high version increased ground clearance from 19in to 30in and overall height from 85in to 95in. Front tire size changed to 6x20s.

1954 International McCormick Farmall Model 300
The Farmall 300 used International's 3.56x4.25in inline four-cylinder engine. At 1050rpm, the engine produced 30.0 drawbar horsepower and 36.0hp on the belt pulley. With its five-speed transmission, transport gear provided 16.1mph, however the Torque Amplifier reduced speeds by one-third to keep the engine operating in the optimum torque range regardless of ground speed.

1954 International McCormick Farmall Model 300
The Farmall 300 high-clearance measured 149in long overall, 13in longer than the standard and row-crop versions. In addition to being 10in higher, it also weighed nearly 700lb more. Rear tread width, adjustable from 48in to 93in on the standards was limited to between 62in and 74in with the high-clearance models.

1956 International Model 600 Industrial
A rare and unusual piece of equipment is this 1956 International Model 600 diesel Industrial. The 600 Series was produced only during 1956 and was succeeded that same year by the 650 Series tractors, although it may be that the first diesel replacement was not until 1959 with the restyled Model 660.

1956 International Model 600 Industrial
International's huge four-cylinder engine displaced 4.50x5.50in bore and stroke. At 1500rpm, it rated 58 drawbar horsepower and 64hp off the belt pulley. In agricultural configuration, a five-speed transmission provided a transport speed of nearly 16mph. It weighed nearly 9,000lb.

1956 International Model 600 Industrial
Production history is incomplete on this Industrial Model 600. It was fitted with a two-cylinder gasoline engine for starting purposes. Electric starter, generator, and lights are included, as are the Torque Amplifier transmission and a version of the Fast Hitch implement attachment system.

Chapter 7

Massey

In 1892, Massey-Harris bought L. D. Sawyer Company, of Hamilton, Ontario, makers of portable steam engines and locomotive-type steamers since the 1860s. Sawyer-Massey expanded production and developed double-cylinder engines producing as much as 35hp. Its hold on Canadian manufacture led it into the gasoline tractor market. Massey and Sawyer separated in 1910. Yet Massey-Harris had investigated gasoline engines and that year, they acquired the Deyo-Macey engine company of Binghampton, New York. Engine production continued there until Massey opened its own factory at Weston, Ontario, in 1916.

In 1917, Massey began importing the Little Bull. The Bull Tractor Company of Minneapolis was one of the first to produce a tractor aimed at the small-acreage farmer. The Little Bull was a tricycle rig that drove only one of its two rear wheels. It had plenty of advertising pull but it was insufficiently tested. Bull sold 3,800 in eight months but its lack of horsepower damaged its reputation.

So in 1915, the Big Bull was introduced, following the

1930 Massey-Harris General Purpose 4WD
Introduced in 1930, it was called the Massey-Harris General Purpose and was rated at 15-22hp and was rated for two 14in plows and drove all four wheels. It was Massey-Harris' first design attempt and, despite a legacy inherited from the Wallis tractors, the company followed its own inclinations and produced a tractor that was very advanced for its time. The steering tie rod comes over the top of the American Bosch U-4 magneto. Oil lubrication was forced by a geared pump and dip pan splash. The clutch runs in oil as well. The entire rear end pivots, much like the articulated four-wheel-drive tractors of today. The tractor used automobile-type steering. It weighed 3,861lb, stood nearly 55in tall, nearly 59in wide, and 119in long. The wheels—without road rims—were standard 8.0x38 steel. Herc Bouris of Sun City, California, restored and owns this GP 4WD.

same design. But it offered variable axle height to level the tractor when plowing. Massey-Harris imported this version. Yet the program seemed doomed. Just as Massey's contract began, Bull lost its manufacturer. Since Bull had no factory of its own, Massey got no tractors. It made other plans, choosing this time to manufacture tractors for itself, under license.

Dent Parrett designed a tractor that he introduced in 1913. Prototypes, powered by Buda engines, were well received. It was his 12-25 that Massey produced in Canada. Parrett used automobile-type steering and abnormally tall front wheels to lessen compaction and increase wheel-bearing life. Massey began production in Weston in 1919; the Parrett 12-25 tractor was known in Canada as the MH-1. The successor MH-2 tractor improved rear-axle lubrication and reduced overall speed.

Massey-Harris modified Parrett's next tractor. But Henry Ford's Fordson was available in Canada and it was advanced over even Massey's new MH-3; price wars with International Harvester in 1922 soon put Parrett out of business in the United States. Ex-

1929 Wallis Model 20-30
In 1927, when the Wallis 20-30 "Certified" was introduced, it was still manufactured by J. I. Case Plow Works. Massey-Harris bought the marketing rights to the tractor in 1928 and sold back the name to J. I. Case Threshing Machine Company. By 1929, when this tractor was first purchased, it was a Massey-Harris tractor. It is owned now by its original purchaser's nephew, Wes Stoelk of Vail, Iowa.

ports to Canada became so vigorous that Massey-Harris simply withdrew for a few years.

Massey's implement lines did well worldwide. Even as the American tractor wars cut the competition by half, Massey again felt that tractors meant income on their balance sheet. Again, Massey chose to import a US-built tractor. Acquiring an existing line saved development time and money.

Enter Mr. H. M. Wallis, President of the J. I. Case Plow Company, and a relative of Jerome Increase Case. By 1912, his Wallis Tractor Company was aware of the need for smaller tractors. The "unit frame" became the patented signature of Wallis tractors. The crankcase sump pan and transmission case were incorporated and enclosed within a U-shaped one-piece steel casting that also served as the

frame for the tractor. This put all the moving parts inside the housing and protected everything except final drive from the elements.

In 1915, Wallis introduced the Cub Junior, the Model J. The J extended the boiler-plate curved frame all the way to the final-drive gears, fully enclosing all the running gear. Despite this additional steel the J weighed about 4,000lb. It was truly a junior tricycle tractor to Wallis' previous products: it weighed one-sixth of the Bear and sold for one-half the Cub.

Wallis sales were handled by J. I. Case Plow Works salesmen out of Racine. In 1922, Wallis introduced the

1929 Wallis Model 20-30
The Wallis tractors introduced the unit-frame design. On the Wallis, this was a U-shaped one-piece tub frame that served as sump pan for the engine and continued on as belly pan for the transmission and differential. (This differed from Ford's Fordson, which mated engine to transmission as load-bearing members without a separate frame.) Wallis tractors used the firm's own four-cylinder engine, with 4.375x5.75in bore and stroke. At 1050rpm, the engine produced 27 drawbar and 35 belt horsepower.

OK, bragging that it produced "America's Foremost Tractor" and trumpeted the OK as "The Measuring Stick of the Tractor Industry." In its University of Nebraska test, it pulled 75 percent of its weight at 90 percent of its maxi-

1930 Massey-Harris General Purpose 4WD
It was a great idea, just about thirty years early. Some operators criticized its lack of power, prompting Massey in later years to offer an optional Hercules six-cylinder engine. But what it really needed were limited-slip differentials on at least the rear axle. In principle, four-wheel-drive should always find solid footing to power through the slippery stuff. In fact, without limited-slip, the tractor was not much more effective than any two-wheel-drive machine.

mum speed. Wallis advertised three-plow power with two-plow weight. Massey-Harris was impressed, and bought Wallis in 1928.

Massey-Harris moved to the front of full-line firms. In 1929, Massey solidified its standing by introducing the 12-20, retaining the U-frame. The 1931 Model MH-25 replaced the earlier 20-30. The U-frame continued into the 1940s as the structural foundation of a succession of well-regarded tractors from Massey. The MH-25 continued in production through the mid-1930s until the Challenger and Pacemaker were introduced as replacements. Wallis engines were improved for these, increasing power and using a four-speed gearbox instead of the MH25's three-speed. The Pacemaker was a standard tractor while the Challenger brought Massey a row-crop version.

Tractor styling arrived at Massey-Harris with the 1938 models. The square corners inherited from Wallis softened with the Streamlined Pacemaker. Wallis's green paint scheme became Massey's red. For 1940, Massey replaced both the standard Pacemaker and row-crop Challenger with "Twin-Power" versions. This feature overrode the 1200rpm governor on drawbar work to provide 1400rpm engine speed for the drive belt. Orchard tractors were available in Twin-Power versions. An optional implement power lift was offered.

In 1930, Massey's engineers took a risk and began to develop a radically new machine. Borrowing nothing from Wallis, Massey introduced its General Purpose model in 1931. Until then, four-wheel-drive successes had been rare.

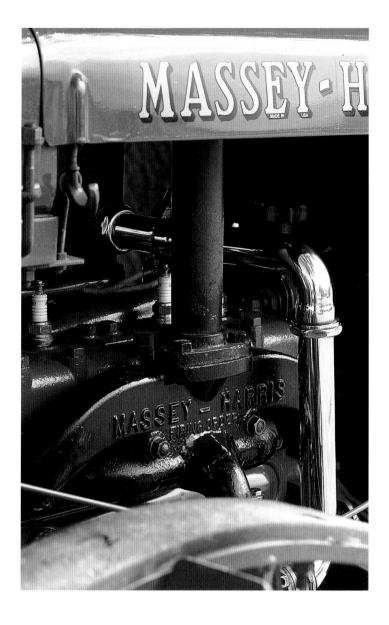

1930 Massey-Harris General Purpose 4WD
Massey outfitted the GP with a Hercules inline four-cylinder engine of 4.09x4.50in bore and stroke. At 1200rpm, the engine produced 15.6 drawbar and 24.8 belt pulley horsepower. When owner Herc Bouris restored the tractor, the air intake pipe had rusted to lace. Never expecting photography he looked around for a suitable, readily obtainable replacement and found plumbing supplies fit perfectly.

1930 Massey-Harris General Purpose 4WD
The ingenious system of bull gears inside pinions inside each wheel hub enabled the GP 4WD to maintain its tall ground clearance—nearly 20in under the sump pan. But the early tractors—this was the eighth manufactured—ran into a variety of problems with lubrication and with steering. Many early tractors had these huge castings replaced.

Complex dual differentials and drive axles that steered produced tractors difficult to maneuver. These problems defeated the virtues for which four-wheel-drive was sought. Massey's GP was basically a row-crop tractor and tread width was available from 48–76in, with nearly 30in of ground clearance.

In 1936, Massey offered the GP for gas or kerosene. Wider tread width was available and rubber tires were offered. Addressing the insufficient power problem, Massey offered a Hercules six-cylinder engine. By 1937, it was out of production because the tractor needed continued development. With the GP-4WD, Massey had jumped about twenty-five years ahead of its times.

While Henry Ford introduced his 9N, Massey brought out its 101 Junior. A small two-plow-rated tractor, it was priced at $895, included a self-starter, battery ignition, rubber tires, and fenders. The company's Twin-Power feature was optional.

But unlike Ford's 9N, the 101 constituted a full-line, with standard, Super, and Senior versions as well as the Junior. As interesting were the MH Senior, Super, and the 201, which all used a six-cylinder Chrysler truck engine modified for Massey. Six cylinders was a legacy from the Hercules attempt with the GP-4WD, and it was quite a luxury in the days when John Deere still stood by its two-cylinder Poppin' Johnnies, and everyone else got along on fours.

Massey believed advantages far outweighed oddity. Chrysler's truck engines advertised 12 billion miles of proven performance. With Chrysler engines, Massey offered the widest ranging dealer service network of any tractor manufacturer.

1947 Massey-Harris Model 30
Massey Harris introduced its Model 30 in late 1946, but its appearance was set years earlier by Massey's designers when they adopted tractor styling. The horizontal louvers of the grille first appeared in the Challengers in 1936. The elliptical wave shape cut-out for the engine began then as well and was continued to greater or lesser extent through the late 1950s.

Massey's small tractors, the MH-101, Junior, and Standard, were supplemented by the MH-81 and General, wartime military versions. Canada's Royal Air Force used the MH-81 to move aircraft. The General was a product of the Cleveland Tractor Company, parents of the Cletrac crawler; it was Cleveland's only effort into the wheeled-tractor market. It was a tricycle high-clearance row-crop. But it remained in Cletrac yellow, never bore Massey's name plate, and appeared in Massey catalogs only two years.

Few innovations caused more competitive "catch up" than Harry Ferguson's system of automatic draft control and increased plow downforce, which had been a great sales advantage to Ford. Massey-Harris' system was called the Depth-O-Matic hydraulic lift and it was first offered as an option around 1950. Yet Massey's hydraulics were not an integral system but an add-on, and the Depth-O-Matic didn't have Ferguson's automatic draft control. The distance between Massey's system and Ferguson's led Massey to an agreement with Harry Ferguson.

As the Ford-Ferguson partnership unraveled, Ferguson introduced his own tractor first in England and then in the

United States. Ferguson advocated mechanization to cut the production costs and worked consistently to keep prices low so farmers could afford the tools necessary to produce more at lower cost. Outside manufacturers had always produced his gray tractors from his designs while he took care of their distribution and sales. Ferguson looked carefully at Massey-Harris. His engineers worked on other projects in England that interested Massey-Harris in Canada. Ferguson met Massey people to discuss Massey's factory in Scotland.

At the last minute, Harry Ferguson vetoed the plans but offered Massey part of his company instead. A visit to Ferguson's shops impressed Massey inspectors with implements being tested.

Massey knew the merger was unlike any other in its history. Every competitor had tried to better Ferguson's system; they all failed and the best they could manage was to steal, copy, or license it. Massey had the opportunity to get not only the system but also the creative mind behind it.

Ferguson had his own ideas, including a new gray tractor, nicknamed the "big Fergie," already in prototype form. Ferguson offered to sell Massey his entire company. Ferguson asked for an honorary "role" as Chairman of the new company. He would handle all engineering matters and projects since Massey's principals were not engineers. Massey's board agreed.

Ferguson sold his company for $16,000,000 worth of Massey stock. He became the largest single stockholder in the resulting company, Massey-Harris-Ferguson. Gray tractors joined the reds.

But misunderstandings followed and the agreement soon broke apart. Massey bought back Ferguson's stock. Massey-Harris-Ferguson continued to operate for some time as two separate competing companies, Ferguson and Massey-Harris. Alanson Harris' name was withdrawn from the logo in 1958 when the company became Massey-Ferguson.

The same year, Sir Edmund Hillary took three track-fitted Fergusons 1,200 miles to the South Pole. In warmer climates, Massey's factory in France introduced a new small tractor, the MF-25, while North American markets got the first MF-85s; the 1958 line ranged from 25hp to 60hp gas or diesel versions. But customers favored larger, more powerful machines and Massey once again sought outside sources. Minneapolis-Moline built a 425ci six-cylinder for Massey, the MF-95.

In 1959, Massey acquired F. Perkins Ltd., a diesel engine manufacturer in Ontario. Massey also acquired an Italian tractor company, Landini, in 1960. It had used diesel engines from its start in 1910. These extremely inefficient engines required enormous displacement to produce little power. Their advantage was that they ran on almost anything flammable.

1947 Massey-Harris Model 30
The Model 30 was very popular, selling nearly 32,500 copies at nearly $2,000 each from 1946 through 1952. Massey fitted it with a Continental Red Seal four-cylinder 3.43x4.375in L-head engine. At 1500rpm, it produced 20.6hp; a feature called Twin Power bypassed the governor for transport or belt work and at 1800rpm, the Continental engine produced 30.1hp. This example is owned by Ted Nelson of Costa Mesa, California.

In late 1959, Ferguson successfully incorporated a limited-slip differential onto a tractor. He had another new tractor—in three sizes—in mind that would utilize the automobile torque converter. But on October 25, 1960, Harry Ferguson died in his morning bath. He was 76.

For decades he blended genius engineering with insecurity and impatience of the genius artist inside him. Not always successfully. His tractors—indeed nearly all his inventions—met his expectations. It was his associates who usually could not.

Chapter 8

White

In February 1962, when White Motor Corporation acquired Cockshutt Farm Equipment, White was the youngster in the farm-equipment business. The company had owned the Oliver Corporation only fifteen months. Within a year, however, it acquired Minneapolis-Moline Power Implement Company. Six years later, White Motor reorganized and established the White Farm Equipment Company in Oakbrook, Illinois. Its pedigree comprised forty-eight companies spanning 160 years.

James Cockshutt's company was founded in 1877. Soon afterwards, Abell Engine & Machine and Universal Tractor, Minneapolis Malleable Iron married Twin City Iron Works. Much later, in 1929, a good year for farm-equipment mergers but bad for world economy, these and others joined Nichols & Shepard Company, Hart-Parr Tractor Works, and Oliver Chilled Plow Company to form Oliver Farm Equipment Company. These grandchildren of the farm-implement evolution reorganized, expanded, and contracted from 1929 to the 1960s. Oliver Farm became Oliver Corporation, Cockshutt became CFE Co. of Canada, and Minneapolis-Moline (MM) incorporated. Between 1944 and 1960, Oliver bought and sold Cleveland Tractor, and in 1951, MM bought B. F. Avery.

Cleveland Tractor began as the Cleveland Motor Plow Company in 1917 founded by Rollin White (of the White Company). White's fortune came from sewing machines. He later developed and produced steam-powered automobiles. The Motor Plow Company experimented with crawler tractors, advertised as "Geared To The Ground."

The crawler's tracks were driven through differentials and planetary gears. To steer the tractor, track brakes were pulled against the main gears to slow one side while the other pulled. White renamed the company Cleveland Tractor Company and by 1918 Cletrac was born.

In 1944, Oliver purchased Cletrac; Oliver crawlers were produced in White's old Cletrac factory. Ironically, of course, White purchased Oliver in 1960, reacquiring Cletrac, which then moved to

1920 Cletrac Hi-Drive Model F 9-16
Roland White's Cleveland Tractor Company produced its first crawlers in 1917. In late 1920, it introduced four versions of the Model F including this "Hi Drive" owned by Mike McGarrity of Pinion Hills, California. Cleveland Tractor continued to manufacture its "Cletrac" crawlers until 1944 when the company was acquired by Oliver Corporation, which produced the Oliver Cletracs until 1960 when White Motors Corporation—the same White family—purchased them. Cletrac had come full circle but was produced for only another couple of years at Charles City, Iowa, home of the Hart-Parr tractors. The little Model F sold for only $850 at the factory in Cleveland. It was not only inexpensive to purchase, it was inexpensive to operate. In its 1922 Nebraska Tests, it was the first tractor to exceed 10hp hours per gallon of fuel, delivering fuel economy of 10.17. At 1600rpm, the crawler engine produced a maximum of 19.6 belt pulley horsepower (though fuel economy suffered).

1913 Hart-Parr Model 30-60
Hart-Parr manufactured its 30-60hp tractor from 1907 through 1918. It was during this period that W. H. Williams, the firm's advertising manager, first used the word "tractor" in advertising. The 30-60, with its huge 10.00x15.00in horizontal two-cylinder, was so dependable it was soon nicknamed "Old Reliable." This 1913 example was restored and is owned by Gary Spitznogle of Wapello, Iowa.

the former Hart-Parr works in Charles City, Iowa, until its own crawler line was dropped in 1963.

The Cockshutt Plow Company of Brantford, Ontario, was established in 1877. Cockshutt added other lines and prospered. By 1924, it marketed Hart-Parr tractors in Canada. In 1928, a year before Oliver acquired Hart-Parr, Cockshutt agreed with Allis-Chalmers to market its tractors in Canada and a 1931 sales brochure showed Allis tractors with Cockshutt nameplates. The arrangement soured. So in 1934, Cockshutt again imported tractors, selling Olivers in Canada through the late 1940s.

Cockshutt considered joint tractor manufacture with Massey-Harris. But it concluded it was too costly and eventually it produced its own. Architectural designer Charlie Brooks styled the slender Model 30 that put Cockshutt on tractor makers' maps.

Thorough testing yielded reliable machines and introduced a landmark improvement: the continuous live power takeoff. Prior to Cockshutt's innovation, when the tractor was slowed, stopped, or the clutch disengaged, the PTO slowed or stopped. Cockshutt made the PTO independent with its own separate clutch. Thus harvesting machinery or other PTO-dependent implements continued running even if the tractor was stationary.

Industrial design came in 1957 when Cockshutt introduced the 500 Series, designed by Raymond Loewy's group. The art-deco stylishness of its first tractors was traded for "neo-purposeful"; bodies were widened, grilles became bolder, and the tractor appeared heavier and more massive.

In 1962, the Oliver Corporation (as White Motors' subsidiary) purchased Cockshutt. Production continued for more than a decade with implements and tractors called Cockshutt, but its identity was absorbed into Oliver.

Hart-Parr Tractors of Charles City, Iowa, came from engineering students Charles Hart and Charles Parr, who designed their first engine while still in college. After graduation, they moved to Charles City, Iowa, where they found backing. By 1902, their first gasoline traction engine was tested in the fields. Their second prototype, tested in 1903, was such a success that a production run of fifteen was completed. In 1907, Hart-Parr introduced its Model 30-60, "Old Reliable," and its advertising manager, W. H.

1917 Moline Universal
The Universal was introduced at $385 by a company in Columbus, Ohio, in 1914. But by the end of 1915, Moline Plow Company in Moline, Illinois, had purchased the company and its machine. The Universal four-cylinder 3.50x5.00in engine was rated as a 9-18hp tractor but produced more than 17 drawbar and 27 pulley horsepower. The Universal offered an electric governor, starter, and headlight as standard equipment. This 1919 Model D belongs to Jim Jonas of Wahoo, Nebraska.

Williams, introduced the first commercial use of the word "tractor" in promoting Old Reliable.

Hart-Parr's machinery was derivative of steam traction engines of the day. Its trademarks were 1,000lb flywheels and 20,000lb tractors. Two-cylinder kerosene engines were

1920 Cletrac Hi-Drive Model F 9-16
Cletrac's Hi-Drive F was driven by a floating roller chain that travelled between the drive gears and the tracks. This was different from other crawler manufacturers—Best and Holt for example—which drove their tracks directly from large toothed final drive sprockets.

1920 Cletrac Hi-Drive Model F 9-16
When Cleveland Tractor introduced its first tractor, it used outside-sourced engines, a Weidley for example. By 1920, Cletrac tractors used Cleveland-manufactured engines. The F was powered by the 3.25x4.50in L-head four-cylinder engine. Tested in Nebraska, the crawler demonstrated exceptional strength: it weighed 1,920lb and in its only forward gear pulled 1,780lb at 2.8mph.

cooled by oil, and make-or-break ignitions systems sparked only on demand as the engine load increased. The tractors were driven and steered by chains and ran at 2.3mph.

Hart-Parr created another "first" in tractor history. Uneducated operators caused equipment failures. Years of this frustration led to instruction programs at various sales branches. Education by mail was available to operators living nowhere near a branch. This idea was adopted by competitors with comparable experiences with repairs caused by operator ignorance or carelessness. Around 1919, however, after much success, Hart and Parr completely withdrew from the company. In 1929, Oliver acquired the firm.

Minneapolis Steel and Machinery Company of Minneapolis, Minnesota, made the Twin City tractors. Twin City Iron Works, a predecessor, was founded in 1889 for heavy steel and iron construction and fabrications. But by 1903 the company (renamed Minneapolis Steel) was producing industrial steam engines. It began manufacturing a German-made gas engine. This enabled it to produce tractors for J. I. Case Threshing Machine.

Ironically, the Twin City's first tractor was built by outsiders. Five successful prototypes allowed Joy-Willson Company to produce several hundred 40-65hp tractors through 1920. It even offered a limited run of crawlers similar to the Northern Holt Company's crawlers with front steering wheels.

Of course, by 1918 Minneapolis Steel had its own smaller tractors and its Twin City 16-30 was advertised as being automobile-like in style and engineering. Its exaggerated length, emphasized by its fully enclosed engine and sides, resembled the rakish sports cars on the market. The tractor's length shortened in the next years.

With considerable experience by this time, Minneapolis Steel introduced its 12-20 Twin City in 1919. It provided a four-cylinder engine with dual camshafts and four-valves

1920 Cletrac Hi-Drive Model F 9-16
Cletrac Model F crawler operators turned a steering wheel to steer the tractor. The wheel braked the inside track while the Cletrac's differential sped up the outside. Turning was crisp and tight on the small Model F although this is not normally the case with differential-steered crawlers. The Model F stands 52in tall to the top of the steering wheel, is 81in long overall, and 43in from fender to fender.

per cylinder, elements incorporated seventy years later on production automobiles. In 1929, the firm joined Moline Plow and Minneapolis Threshing Company to create Minneapolis-Moline Power Implement Company.

The Moline Plow Company was formed in 1870, the result of observant study of the farm implement business by two partners, Robert Swan and Henry Candee, who came together in 1852 to purchase a manufacturing operation in

1922 Minneapolis Threshing Machine Model 35-70
When it was introduced in 1920, Minneapolis Threshing Machine Company rated this as a 40-80hp tractor. Nebraska Test results forced a designation change and as a 35-70, its performance was impeccable. According to the best information available, this example, number 4018, was manufactured early in 1922. It was restored and is owned by Herc Bouris of Sun City, California.

1922 Minneapolis Threshing Machine Model 35-70
Overall, it stands 11ft tall. It measures 10ft wide and 17ft, 6in long. Its rear drive wheels are 30in wide by 7ft, 3in tall; the front wheels are 14x42. It weighs 24,000lb. The cabin is tall enough for a six-footer to stand at full height. This machine was one of the great prairie sodbusters, capable of drawing a dozen 12in or 14in plows through the ground at 2mph—nearly 4.5 acres per hour.

1922 Minneapolis Threshing Machine Model 35-70
The Model 35-70 is an engineering marvel for 1920. Two separate oilers provide lubrication: the front one with twelve stations feeds the engine and is chain driven off the fuel pump shaft, which is chain driven off the camshaft. The second oiler, behind the front one, also has twelve stations, to lubricate the running gear and is belt operated off a special gear that is only driven when the tractor is in motion. The galvanized tank behind the fan belt houses the governor.

Moline, Illinois. The Universal Tractor Manufacturing Company was an offshoot of Ohio Carriage Manufacturing in Columbus, Ohio, which had produced one small prototype for them in 1914. The Universal worked well. By 1916, 450 had sold at $385. It used a Reliable two-cylinder engine and was conceived more as a motor cultivator than as a tractor.

In November 1915, Moline Plow bought Universal, as the company introduced its Model D, a larger four-cylinder machine. The two-cylinder Universal continued until 1917 when the D was offered with a Moline Plow-built engine. Universal buyers in 1918 got the first tractor providing an electric self-starter and a headlight as standard equipment. By 1923, as the postwar depression dug in, Moline Implement dropped the Universal Motor Plow. In mid-May 1929, it merged with Minneapolis Threshing Machine and Minneapolis Steel and Machinery. The worldwide depression that broke many competitors brought these three firms together as Minneapolis-Moline Power Implement Company.

The conglomerate's first products emptied existing stock. Twin City tractors of the Minneapolis Threshing Machine line continued until 1931 when M-M introduced its own machine. The new Twin City line was well-engineered, thoroughly tested, and fully developed. Its water pumps, variable fuel carburetion, dual-system air cleaners, and pressure oil lubrication pumps let Minneapolis-Moline

1922 Minneapolis Threshing Machine Model 35-70
The red tank above the radiator holds gasoline for starting. Once running, the 35-70 operates on kerosene contained in a 60-gallon tank below the cabin floor. Gasoline is gravity fed to the fuel reservoir in the cabin. Kerosene is pumped. The spinning flywheel is part of the drive-gear clutch.

1922 Minneapolis Threshing Machine Model 35-70
The Minneapolis-built engine is made up of massive pieces. The horizontal four-cylinder engine displaces 7.25x9.00in bore and stroke. A galvanized tin air-intake stack feeds the Kingston Model E carburetor. The large brass tube far left is fuel intake, the smaller one hidden behind the air intake is water-injection feed—to control pre-ignition. Upper right is the throttle. The tall gray lever is the main drive clutch lever.

claim "Three Extra Years is the Reputation of All Twin City Tractors."

Minneapolis-Moline introduced an original design in 1931 with its Model M (Universal), a row-crop general-purpose tricycle. In 1934, it introduced the new Universal and the Standard Model J. Both used an F-head four-cylinder engine (intake above, exhaust below), which prolonged valve life and increased power.

Styling was prevalent in the industry by 1938 and bright colors helped sell tractors. To introduce its Prairie Gold tractors, M-M showed off its new styled machinery to 12,000 invited guests. The striking Model UDLX (U Deluxe) joined the lineup of "Vision-lined" tractors. The show startled the visitors. M-M intended its UDLX "Comfortractor" as a dual-purpose machine to work the farm and combat the automobile. With its fully enclosed cab and strong resemblance to an automobile, the farmer could work the field all day and then—in fifth gear at 40 mph—take his family to town for the church social. But introduced at $1,900, it cost nearly $1,000 more than a standard U, and was too ostentatious for some farmers. Still, with its heater and radio, the UDLX made many owners feel a sense of accomplishment. About 150 sold and most were heavily used year round, even transporting and operating corn shellers in the winter throughout the Midwest.

M-M innovations created another legendary machine. The UTX tractor, introduced in late 1938, was an all-wheel-drive machine for the military and was first tested by the Minnesota National Guard. Produced during World War II, it took its name from a regular character in the "Popeye The Sailorman" cartoons: "Jeep" could do anything and knew everything. UTX-conversions fit the bill

1922 Minneapolis Threshing Machine Model 35-70
The belt pulley, off which 74hp was generated in Nebraska Tests, also houses its clutch and the "crank" starting ratchet. The "crank" is actually a nearly 4ft long lever with a ratchet hub. The bevel-cut teeth inside the belt pulley kick the "crank" back—disengaging it—in case of backfire. With 74hp on tap, if it didn't release, it would catapult the operator a country mile up and over the drive wheels.

1922 Minneapolis Threshing Machine Model 35-70
The Minneapolis offered a spacious cab with plenty to treat the eye and plenty of noise to assault the ears. The tall lever at left is the gearshift, with two speeds forward. The brass can at top left is the fuel reservoir and sight gauge. Kerosene is constantly pumped up into it and overflow returns to the main tank. The lever near the oil can top right is the belt pulley clutch and below the steering wheel is a brake. No operator seat is fitted.

and a Guardsman at Camp Ripley nicknamed one of the first ones. The name stuck. Thousands were built.

However, in 1969, Minneapolis-Moline was acquired with Oliver Farm Equipment and Cockshutt Farm Equipment of Canada to become the farm equipment division of White. Forty years after their merger to avoid financial disaster and economic ruin, farm-implement history repeated itself: Cockshutt, M-M, and Oliver, hobbled by overproduction and sagging farm incomes, joined forces to combine resources, eliminate duplication, concentrate production, and stay in business.

More than a century earlier, the chilled plow was perfected in a process invented by James Oliver. In 1855,

Oliver cooled freshly cast iron with a stream of water. During his process, he annealed—or glazed—the plowing surface to polish it and make it resistant to rust while improving its ability to scour. The slow, water-bath cooling increased the iron's strength without making it brittle.

In the spring of 1929, even as Minneapolis-Moline Power Implement was coming together up in the Twin Cities, Oliver Farm Equipment was forged from the assembly of James Oliver's Chilled Plow Company, Hart-Parr Company, Nichols & Shepard Company and American Seeding Machine Company. Oliver expanded into four di-

1927 Oliver Hart-Parr Model 28-50
The Model 28-50 was conservatively rated by Hart-Parr. When it was tested at the University of Nebraska in August 1927, it far exceeded its manufacturers' advertising specifications. Run at 850rpm, the drawbar horsepower was actually 43.6hp while the engine produced a maximum of 64.6hp on the belt pulley.

visions; the former Oliver represented only plow and tillage equipment; Nichols & Shepard contributed its Red River Special threshers; Superior Seeding provided seed planting and fertilizing machinery; and Oliver's Hart-Parr division manufactured and marketed the tractors.

This new Oliver-Hart-Parr line was introduced in 1930 with an upright, longitudinally mounted four-cylinder engine. Orchard and rice field versions were offered from the start, and in 1931 an industrial version was added. Hart-Parr adopted rubber tires and in 1932, pneumatic tires replaced the hard rubber.

These tractors boasted high-compression power and streamlined beauty from the first production versions.

1927 Oliver Hart-Parr Model 28-50
The 28-50 was only available through 1928, selling for $2,085. It was 135in long, 88in wide, and 64.5 in high without an optional cab. Wheels measured 7x28 front and 14x51 steel rear. This late-series Hart-Parr was restored and is owned by Mike McGarrity of Pinion Hills, California.

1927 Oliver Hart-Parr Model 28-50
The firm set two of its horizontal two-cylinder 5.75x6.50in engines side-by-side—engine from the Model 12-24—to power the big 28-50. Valve clearance—in fact engine maintenance of all types—was critical to the long life of all tractors with running gear exposed to the dust. Hart-Parr went so far as casting the warnings into the engine.

1927 Oliver Hart-Parr Model 28-50
Hart-Parr's Model 28-50 was the last large tractor the firm introduced before it was acquired by Oliver Farm Equipment Company in 1929. It was also the most powerful produced after the product reorganization in 1918, introducing "The new Hart-Parr line," all of which were manufactured at the firm's Charles City, Iowa, plant.

Called the Oliver Model 70, these used a small-displacement automobile-type six-cylinder engine. Steering brakes provided an 8ft turning radius. All engine and implement controls were within finger reach of the automobile-style steering wheel.

Its clean bodywork was most striking. The raked-back grille was crowned with the Oliver Hart-Parr logo. Nothing protruded from the sides or top except a tall exhaust pipe and row after row of cooling louvers. Electric lights, starter, and battery were included by 1938. Standard and orchard versions adopted the same stylish appearance.

Oliver involved the farmers by sponsoring a color contest at state fairs to vote on a new model's official color combination. Tractors painted in all the candidate colors were displayed—and sold—around the country. Meadow green with clover white trim was the winner.

In 1960, Oliver celebrated its twenty-fifth anniversary with six-cylinder engines. Development and technology had taken a 201ci six producing 28hp at 1500rpm up to 76hp at 2000rpm from a 265ci six-cylinder. Within days of this celebration, Oliver Farm Machinery Company was sold to White Motor Corporation.

In 1962, this Oliver Division purchased Cockshutt Farm Equipment of Canada. In 1969, when White acquired Minneapolis-Moline Power Implements, a new company was born: White Farm Equipment. Mergers through the 1960s assembled ideas, patents, experiments, and profits from four dozen farm machinery makers, and consolidated vast history under one corporate roof. White continued tractor manufacture in Charles City, home nearly a century earlier to its ancestors, Charles Hart and Charles Parr.

1927 Oliver Hart-Parr Model 28-50
With the Robert Bosch ZU4 magneto squeezed alongside, any repair or restoration work on the Madison-Kipp twelve-station oiler would have required great patience and small hands. Hart-Parr used the Schebler Model D carburetor, which managed the fuel switch from gasoline for starting to distillates for operation.

1927 Oliver Hart-Parr Model 28-50
In its Nebraska Tests, the tractor weighed 10,394lb. Fitted with a two-speed forward transmission, the strongest pulling power came at rated speed in low gear when it dragged 7,347lb at 2.2mph. The tractor was rated for five plows of 14-16in bottoms.

1930 Oliver Hart-Parr Model 28-44
The Oliver Hart-Parr 28-44 was originally known as the Hart-Parr 3-5 Plow Tractor, and was introduced in mid-1930. An advertisement in *Implement Record* announced that one-sixth of its total weight of 5,565lb was in steel forgings. This is easy to believe considering the size of the U-shaped housing comprising engine sump, transmission belly pan, and differential case.

1930 Oliver Hart-Parr Model 28-44
Sometime late in 1930, this tractor adopted two new names. It became the Oliver 28-44 when the Hart-Parr name began to disappear from all Oliver Farm Equipment tractors. Almost simultaneously it became known as the Oliver 90 at the point when Oliver's model numbers related less to performance characteristics than marketing needs.

1930 Oliver Hart-Parr Model 28-44
Hart-Parr's four-cylinder 4.75x6.25in engine was sparked by an American Bosch U4 magneto and fed by an Ensign Model K carburetor and a Vortex air cleaner. Engine output was 28.4 drawbar horsepower and 49.0 horsepower off the belt pulley at 1125rpm. The cylinder head was cast in chrome-nickel iron; removable cylinder sleeves and pistons were cast from nickel-alloy iron to resist wear. Timken roller bearings and SKF ball bearings were used throughout.

1937 Minneapolis-Moline Twin City Model JT-O
Minneapolis-Moline's Twin City Model J Series tractors were produced from 1936 through 1938. The TC line was well conceived, with variable fuel carburetion, dual-system air cleaners, pressure oil lubrication pumps, and water pump. In appearance, this 1937 JT-Orchard (JT-O) featured more sheet metal mounted to protect fragile tree branches from damage. It is owned by Walter and Bruce Keller of Kaukauna, Wisconsin.

1939 Minneapolis-Moline UDLX Comfortractor
Fifth gear in the Minneapolis-Moline Model UDLX "Comfortractor" was road gear and for MM, this meant 40mph. The UDLX was conceived as the tractor to combat the car: its cab accommodated two or three people, and after a week working in the fields, the farmer and his family could use the UDLX to get into town for socials, supper, the cinema, or church on Sunday.

1939 Minneapolis-Moline UDLX Comfortractor
Minneapolis-Moline introduced high-compression engines to the farm in 1935. Its engines ran with 5.25:1 compression while the competition still used engines of 4.0:1 or 4.1:1. The higher compression produced more horsepower and returned better fuel economy but required higher octane fuel: gasoline. Roger Mohr of Vail, Iowa, owns this 1939 model, other examples of which his father sold as the local Minne-Mo agent.

1939 Minneapolis-Moline UDLX Comfortractor
A survey indicated farmer/operators wanted tractors with cabs. MM responded with the Comfortractor, which lived up to its name. Such amenities as a heater, radio, a clock imbedded in the rearview mirror, and a fold-up jump seat were among other features included in the $1,900 price. To some buyers it was an indication of accomplishment. To others—who couldn't afford it—it was showy. Fewer than 150 sold.

1947 Oliver Standard 70

Oliver introduced its Standard 70 in 1935. The model name referred to its front end configuration and the gasoline octane rating required to operate the tractor. Prior to this series, radiator grilles bore not only Oliver's name but also Hart-Parr, which Oliver had acquired in 1929. With this series, streamlined styling also appeared.

1947 Oliver Standard 70

Electric starter and headlights were standard equipment on most tractors by the mid-1940s. Instrument panels—and overall design— were really improved with the arrival of industrial design. For power, Oliver used its own six-cylinder engine with 3.125x4.375in bore and stroke. It produced 22.7 drawbar and 30.4 belt pulley horsepower at 1500rpm.

Next page
1947 Oliver Standard 70

Oliver advertised and promoted its streamlined styling. The gently curved radiator grille, the contrasting color treatments, and the sleek cooling louvers along the engine covers were meant to suggest that Oliver's tractors were as up-to-date as any machine available—including any automobile the farmer might fancy. This sleek 1947 Standard 70 is owned by John Jonas of Wahoo, Nebraska.

1947 Cockshutt Model 30
Cockshutt introduced the independent continuously running power takeoff (PTO), which kept power to implements even while the tractor was not moving forward. This simply required a separate driveshaft and clutch system, not unlike the system used for belt pulleys. But successfully snaking it through the differential had proven a difficult challenge. Jeff Gravert of Central City, Nebraska, restored and owns this 1947 Model 30.

1947 Cockshutt Model 30
Cockshutt's Model 30 was tested at the University of Nebraska in May 1947. Cockshutt's Buda-built four-cylinder engine displaced 3.43x4.125in bore and stroke and at 1650rpm, produced 21.7 drawbar horsepower and 30.3hp on the pulley. Jeff's father, Carroll Gravert, operates a tractor restoration shop in Central City, Nebraska, where Jeff learned his skills.

1948 Minneapolis-Moline Model U
Dale Gerken's 1948 Minneapolis-Moline Model U almost gets lost in the late harvest corn in central Iowa. Minneapolis-Moline called their corporate logo color Prairie Gold and its flaxen color applied equally to wheat and corn and many other crops near harvest. It was a considerable change from the gray of MM's Twin City line—and of many other makers—barely a decade earlier.

1948 Minneapolis-Moline Model U
Model U tractors were rated for three or four 14in plows and were fitted with five-speed transmissions, allowing transport gear top speed of between 15 and 20mph, depending on options. Standard equipment front tires were 6.00x16 while rears were 13x38s. The tractor weighed around 6,000lb and sold new for $1,800.

1948 Minneapolis-Moline Model U
Minneapolis-Moline built its own engines for the Model U. This vertical, inline four-cylinder measured 4.25in bore and 5.00in stroke. At 1275rpm, the engine produced 26.8 drawbar horsepower and 33.4hp on the belt pulley. MM used Delco-Remy electrics and Ensign's Model KGL carburetor.

1954 Minneapolis-Moline Model UTU LPG
Minneapolis' Model UTU was introduced in 1939, although propane wasn't widely available as an option until late 1953. In 1954, tests at the University of Nebraska indicated that the LPG engine compression was 8.0:1 compared with 6.3:1 for gasoline engines. Bore and stroke were standard U engine specs: 4.25x5.00in The propane engine produced 35.7hp on the drawbar compared to 33.6hp for the gasoline engine. This unrestored 1954 model is part of the collection at the Antique Gas and Steam Engine Museum of Vista, California.

Chapter 9
Orphans

Orphan tractors were machines that were not absorbed or acquired by surviving major producers. They may have had intrinsic value but their production ended because no profitable market existed.

For all the tractor makes that were "adopted," many more simply went out of business. In some cases these makers produced only one model for a year or two, such as Graham-Bradley's Model 32. Or they survived decades of success with other products and then failed to survive a technology changeover, such as Russell & Company.

The orphans came from major national firms and minor local ones. They were equally the results of manufactured products that lacked quality and of quality manufacturers that lacked suitable products.

1921 Samson Model M
Samson Tractor Company introduced its Model M in 1919. Samson was owned by General Motors Corporation, which had purchased the Stockton, California, based maker in 1917. GM's goal was to compete with Ford's Fordson but Samson's Sieve Grip—first shown in 1916—was no match; it was too large and too expensive. GM engineer Arthur Mason designed a replacement—the Model M—but GM president Will Durant had other ideas. By 1921, it was GM stockholders' idea to quit producing tractors all together. Samson specified the Simms Model K4 magneto and it used Kingston's Model L-2 carburetor. It started on gasoline and switched to kerosene once operating temperature was reached. The M tractor was manufactured at the Wisconsin plant of Janesville Machine Company, which GM purchased in 1918. GM soon moved all Samson production to Wisconsin. This restored 1921 Model M, number 24892, is owned by Lee Dyal of Placentia, California.

1911 Fairbanks Morse Model 15-30
Fairbanks Morse began experiments with gasoline engines by 1893, and over the next fifteen years, produced self-powered railroad work cars. Tractor experiments began in 1910. The Fairbanks Morse Model 15-30 used the screen system for engine cooling. Engine-heated water was fed by a belt-driven centrifugal water pump to the top of the screen where gravity and evaporation did the rest. The Model 15-30 weighed about 16,000lb. Fairbanks Morse tractor production ended in 1914.

1911 Fairbanks Morse Model 15-30

Power in those days was inefficiently achieved. The huge tractor derived its 15 drawbar and 30 belt pulley horsepower from one cylinder with 10.50in bore and an 18.00in stroke running at 250rpm. Total displacement was 1557ci. The firm continued producing stationary engines until 1918. Afterwards the company name appeared only on weigh scales. This rare tractor is owned by the Antique Gas and Steam Engine Museum of Vista, California.

1915 Russell "Giant" 60hp

Russell & Company of Massillon, Ohio, began producing gasoline-engine tractors in 1909. A Model 40-80, known as the "Giant," was introduced in 1913 but was renamed after its Nebraska Tests. Fitted with its own transverse-mounted upright four-cylinder engine of 8.00x10.00in bore and stroke, the huge engine ran at 525rpm and produced 43.5 drawbar and 66.1 belt horsepower on kerosene. The "Giant" weighed 23,380lb. This example was restored and is owned by the Agricultural Machinery Collection at the University of California at Davis.

1916 Happy Farmer Model F 12-24
The Happy Farmer Tractor Company produced its first tractors not from its own facilities in Minneapolis but from Wilcox Motor Company in 1916, makers of the Sta-Rite engines. This was common procedure for the time; tractor manufacturers were often design and marketing offices only. Fancy drawings tantalized investors who bought stocks in the company, not tractors. But the Happy Farmer was legitimate, albeit short-lived.

1916 Happy Farmer Model F 12-24
Maneuverability was a strong suit of the Happy Farmer. Its nearly 345-degree steering gear atop its single front wheel allowed for tight turning. Turning still required plenty of forethought: from full left to full right required more than twenty turns of the crank-style wheel. One forward speed provided a maximum of 2.5mph.

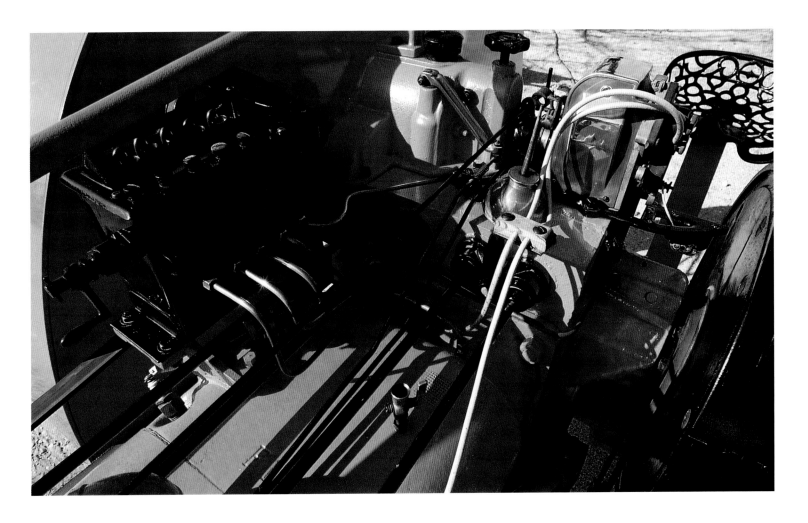

1916 Happy Farmer Model F 12-24
The five-station oiler and the Atwater-Kent K-3 magneto sit atop the horizontal two-cylinder engine. After the first 500 were produced, LaCrosse Implement Company took over manufacture in 1916. The tricycle Model F sold for $1,075. In 1921, Oshkosh Tractor Company bought out LaCrosse but then ran out of money and the LaCrosse Happy Farmer tractor disappeared.

1916 Happy Farmer Model F 12-24
An ingenious induction system pulled radiator-heated air into the Kingston dual-fuel carburetor (gasoline or kerosene). This Model 12-24 used a horizontal two-cylinder engine of 6.00x7.00in bore and stroke. At 900rpm, the engine produced 17.8 drawbar horsepower and 24.2hp off the pulley. Exhaust was directed through the large-diameter forward frame tube. It blew right out the front end—right up and back into the operator's face.

1916 Happy Farmer Model F 12-24
D. M. Hartsough designed the earlier Happy Farmer 8-16, in tricycle and four-wheel configurations. Hartsough had previously designed and produced the Gas Traction Company's Big Four 30 in 1910 and then the Bull Tractor Company's Little Bull and Big Bull in 1914 and 1915. This rare tricycle Model 12-24 was restored and is owned by Mike McGarrity of Pinion Hills, California.

1918 Samson Sieve Grip Model S-25hp
The Samson Iron Works was purchased by General Motors Corporation in an effort to compete against crosstown rival Henry Ford. By 1918, Samson's Model S-25 Sieve Grip bore GMC badges everywhere. The four-cylinder engine with 4.75x6.00in bore and stroke produced 12 drawbar and 25 pulley horsepower and a pronounced bark from its short, fat exhaust pipe.

1918 Samson Sieve Grip Model S-25hp
Its unusual wheels created its name—Sieve Grip—and they provided much more traction than first glance would suggest. The 5,000lb tractor was designed and manufactured in Stockton, California. It was configured for the same river-bottom soil that spawned Best and Holt crawlers. This 1918 example is owned by Fred Heidrick of Woodland, California.

1919 COD Model B
The COD Tractor Company moved from Crookston, Minnesota, into Minneapolis in 1919, the year it introduced its Model B tractor. Rated at 13-25hp, the Model B was essentially an update of COD's first tractor, which was designed and built in 1909.

1919 COD Model B
Belt pulleys ran off the flywheel to operate the oil pump, fuel pump, and water pump. The Model B was rated for three 14in plows or one 24in thresher. The tractor was 156in long overall, 78in wide, and 76in tall. It drove on 70in diameter rear wheels and weighed 6,300lb.

1919 COD Model B
The Model B sold for $1,395 in 1919, its last year of production. Manufacturing rights for this tractor were sold to Minneapolis Threshing Machine Company and that firm's 1920 Model 12-25 bears strong resemblance. The significance of the initials COD has long been lost, and few of these tractors remain in existence. This one is owned by Fred Heidrick.

1919 COD Model B
Albert Espe was a contemporary of D. M. Hartsough, who designed the Big Four 30 and the LaCrosse Happy Farmer 12-24. Espe started a machine shop and foundry in Crookston, Minnesota, in 1898 when he was 26. He made his first tractor in 1907 and organized the Crookston Manufacturing Company to manufacture it in 1910. Espe also designed tractors for J. B. Bartholomew at the Avery Tractor Company.

1919 COD Model B
COD used its own engine, a two-cylinder L-head design. It measured 6.50in bore and 7.00in stroke, and at 550rpm it developed 13hp on the drawbar and 25hp on the belt pulley. Modern wiring, jury-rigged during an interruption of restoration-in-progress, criss-crosses the engine and connects it to the governor (upper right).

1919 Fageol Model 9-18
Fageol Motor Company of Oakland, California, introduced its Fageol tractors in 1917 with an 8-12hp standard configuration four-wheel machine. In its 1923 sales brochures, the manufacturer was listed as the Great Western Motor Company in San Jose. Great Western appears to have taken over manufacture and produced trucks and buses and even four-cylinder upright motorboat motors.

1919 Fageol Model 9-18
Fageol's brochure attempted to explain its unusual drive wheels: "Drive a wedge, the shape of a Fageol 'leg' or grouser, six inches or less into the ground. It will sustain your weight, be easy to pull out and yet defy you to drag it sideways. This is the Fageol Traction Principle."

1919 Fageol Model 9-18
The Model 9-18 weighed 3,500lb. Overall, it measured 119in long, 55in wide, and 52in tall. It sold new for $1,500 from the Oakland factory. The grousers or "legs" each were 10in long, 2.25in across the face. Thirty-two "legs" made up a wheel and Fageol calculated that with penetration of 6in, their unusual wheels provided 170sq-in of ground contact.

1919 Fageol Model 9-18
The 1919 Model 9-18 was a two 14in plow-rated tractor, which used a Lycoming 3.75x5.00in inline four-cylinder engine. At 1200rpm the conservatively rated Lycoming produced 22.5 drawbar horsepower and 35hp on the belt pulley. Lycoming used a five-main-bearing crankshaft. For ignition, Fageol used either a Dixie or a Splitdorf high-tension magneto with automatic impulse starter.

1920 Avery Model C
The Avery Company was founded in Galesburg, Illinois, in 1874 by brothers Robert and Cyrus Avery. The company produced primarily planters and cultivators through 1884, when it relocated to Peoria. Three years later, its first steam traction engines appeared. An early attempt with a gasoline tractor failed but by 1911, Avery produced a successful machine. This low-slung Model C was introduced in 1920.

1920 Avery Model C
Albert Espe had designed the COD tractors and was hired by Avery's president J. B. Bartholomew in 1912. Espe was to design a small tractor to pull Avery away from the giant machines it had built while the founding brothers were still alive. The Model C was as near as Avery and Espe got to an orchard-configuration tractor.

1920 Avery Model C
A clever innovation used the six-cylinder engine's exhaust to clear the drive gears. On both sides of the tractor, exhaust pipes exited just below the drive pinions with enough force to blast away the dust that might otherwise have worn out the gears. Avery must have considered the effect of the heat less harmful than the dust.

1920 Avery Model C
The 1920 Model C used Avery's own six-cylinder engine, the first six ever tested at the University of Nebraska. With bore and stroke of 3.00x4.00in, the L-head inline six produced 8.7 drawbar and 14.0 belt pulley horsepower at 1250rpm. Early in 1924 Avery went bankrupt, reorganizing later that year. The Avery Power Machinery Company continued until 1931 when the Depression sunk it. One last attempt in the late 1930s produced the Ro-Trak but this was doomed by material shortages in World War II.

1921 Huber Super Four 15-30
Edward Huber began manufacturing hay rakes in Marion, Ohio, in 1865. Some thirty years later he purchased patents from Van Duzen Company in Cincinnati and produced about thirty single-cylinder gasoline tractors but without much success. In 1914, he tried again—and failed again. It was not until the introduction of his Light Four in 1917 that he met success. This Super Four 15-30 was introduced in 1921.

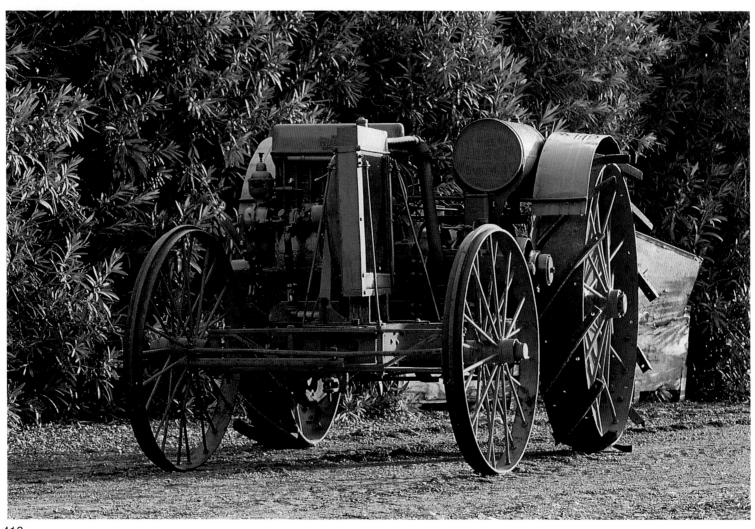

1921 Huber Super Four 15-30
The 1921 Super Four 15-30 was quickly recognized by its large front wheels that were carried over from the first Light Four models. The Super Four remained in production into 1924 when it was modified and its power increased to 18hp on the drawbar and 36hp off the belt pulley. The tractor went out of production in 1925 while the entire Light Four line remained in production only until 1928.

1921 Huber Super Four 15-30
Huber used an inline four-cylinder engine produced by Midwest Engine Company. With 4.50x6.00in bore and stroke, the engine produced 26.9 drawbar horsepower and 44.7 belt pulley horsepower at 1000rpm. Kingston made both the Model L carburetor and the Model LD4 magneto. Its top speed—in second gear—was 4.2mph. The 6,090lb tractor pulled 3,645lb in low gear.

Next page
1921 Huber Super Four 15-30
Huber tractors evolved into standard and row-crop models known as the Modern Tractor series. These were introduced in 1929 to replace the Light Four lineup. A styled tractor was produced in late 1937 and an orchard model also appeared. But these were doomed—as were tractors of other makers—by World War II. Huber emerged from the war but continued to manufacture only construction equipment. This 1921 Super Four was one of Fred Heidrick's first restorations and is still part of his collection.

1921 Samson Model M
Its resemblance to Henry Ford's Fordson was more than accidental. Samson's earlier Sieve Grip configuration with its unusual wheels was fine for spongy soil and its low profile was beneficial to orchard work in California. But it was not appropriate to the rest of the United States. A GM engineer designed Samson's Fordson rival, which was first produced in May 1919 at a rate of ten units per day.

1921 Samson Model M
Samson's Model M used Samson's own four-cylinder L-head engine with 4.00x5.50in bore and stroke. At 1100rpm the engine produced 11.5 drawbar and 19.0 pulley horsepower in a tractor that weighed 3,300lb. This compared favorably with the 2,700lb Fordson whose 4.00x5.00in inline four produced only 9.3 drawbar and 18.2 belt pulley horsepower at 1100rpm.

1921 Samson Model M
The Samson offered two speeds forward, with a top speed of 3.2mph in high gear (lever up to the left). The Fordson used a three-speed gearbox and top speed was 6.8mph. The Model M was designed by GM engineer Arthur Mason. It did not suit founder William Durant, who had his tractor division pick up Jim Dandy's motor cultivator and sell it as the Samson Iron Horse, controlled by horse-bridle-type reins.

1921 Union Sure-Grip Model D 12-25
Union Tool produced its Sure-Grip at its Torrance, California, factory only in 1921 and 1922. The low-slung rear operator's position suggested orchard applications were intended in designing this machine. Union manufactured its own 4.75x6.00in engine which it rated as 12 drawbar and 25 belt pulley horsepower. The crawler weighed nearly 10,000lb including its freewheeling, castor-style front wheel. This 1921 machine is believed to be the only Sure-Grip remaining and is now part of the Agricultural Machinery Collection at the University of California at Davis.

1923 Allwork II Model F
"Light in weight, with big surplus of power for general farming and orchards as well as for all kinds of belt work" was how the Electric Wheel Company (EWC) of Quincy, Illinois, described its Allwork II Model F 14-28 tractor, introduced in 1920. The company name referred to its patented wheel-manufacturing technique, not to electric-powered tractors.

1923 Allwork II Model F
Electric Wheel produced tractors beginning in 1912 from experiments begun in 1908. Tractor production continued through 1930. The Allwork II Model F provided a three-speed transmission with a top speed of 3.75mph although the company specified "plowing speed" was second gear, at 2.5mph in direct drive at 900rpm.

1923 Allwork II Model F
The Model F was powered by EWC's own four-cylinder 4.75x6.00in engine. At 900rpm the engine produced 14 drawbar and 28 belt pulley horsepower. It used a Kingston carburetor and American Bosch magneto. The 4,800lb tractor was rated for three 14in plows or a 24in thresher. This is part of Fred Heidrick's collection.

1923 Allwork II Model F
The Allwork II fitted 12x38 rear electrically welded steel wheels, and 5.00x24 fronts. Overall the tractor measured 120in long, 52in wide as well as 52in tall on a 75in wheelbase. The front axle used automobile-type steering. The Model F sold for $1,475. In March 1957, EWC was acquired by Firestone Tire and Rubber company.

1938 Graham-Bradley General Purpose
The Graham-Bradley General Purpose tractor was produced by Joseph, Robert, and Ray Graham. In 1928, they acquired the Paige Automobile company in Detroit. The firm's reputation for speedy and stylish cars influenced the styling of their tractor introduced in 1938.

The tractor used their automobile's inline six-cylinder 3.25x4.375in engine. With Delco-Remy electricals and a Schebler carburetor, the Graham-Bradley produced 19.1 drawbar and 28.3 belt horsepower at 1500rpm.

1938 Graham-Bradley General Purpose
This 4,955lb tractor was sold exclusively through Sears Roebuck & Co. catalogs in 1938 and 1939 in both row-crop and standard configurations. Production ended during World War II and although it planned to resume production after the war, Graham-Paige merged with Henry Kaiser and produced Kaiser and Frazer automobiles instead. This 1939 model is owned by Dale Gerken of Fort Dodge, Iowa.

Index

FARM TRACTORS: A LIVING HISTORY

PHOTO INDEX

Note that when several of the models listed consectutively had the same owner, the owner was listed after the last listing. The first two Allis-Chalmers models are an example. Both the Model E and Model UC are owned by Conrad Shoessler.

Allis-Chalmers Model E, 74-77
Allis-Chalmers Model UC, 126-129
   owner: Conrad Schoessler
Allis-Chalmers Model U, 118-121
   owner: Tractor Testing Museum, University of Nebraska
Allis-Chalmers WD-45, 86-89
   owners: Clyde & Jeannette McCollough
Best 110hp steamer, 15
   owner: Edith Heidrick
Best Model 25, 30-33
   owner: Jerry Clark
Case Model L, 136-139
   owner: Conrad Schoessler
Case Model VACs, 174-177
   owners: Raynard & Ruth Schmidt
Caterpillar D6, 82-85
Caterpillar D4, 107-108
Caterpillar Model 15, 53-55
   owners: Frank & Evelyn Bettencourt
Caterpillar diesel "Old Tusko", 94-96
   owner: Allan Anderson
Eagle C-6, 112-115
Field Marshall Diesel, 90-93
   owners: Walter, Lois, Bruce, & Judy Keller
Farmall F-30 high clearance, 130-133
Farmall M with cotton picker, 225
Farmall MDV, 78-81
   owner: Bob Pollock
Farmall F-30, 160-162
   owner: Ray Pollock
Ford 8N-C pump, 190
Ford 8N-AN, 145
Ford Dexta, 104-106
Ford Golden Jubilee, 184-186
Ford 600 Moto-Tug, 198-200
Ford Model 650 with ARPS Half-Track, 192-193
Ford Model 801, 194-197
   owners: Palmer & Harriet Fossum
Ford Major & Dexta, 97
   owner: Carlton Sather
Ford Model 501 Offset Diesel, 102-103
Ford-Ferguson 2N, 187-189
Ford 971 LPG, 140-146
   owners: Dwight & Katy Emstrom
Ford Model T tractor conversion, 46-49
   owners: Don & Patty Dougherty
Ford (not Henry) Model B, 62-67
   owner: Tractor Testing Museum, University of Nebraska
Fordson with crawler kit, 208-209
   owner: Edith Heidrick
Fordson with corn picker, 34-36
   owner: Kermit Wilke
Fordson, 60-61, title page
   owners: Palmer & Harriet Fossum
Gambles Model 30 (Cockshutt), cover, 220-224
   owners: Raynard & Ruth Schmidt, Vail
Gibson Model E, 216-217
   owner: Paul Brecheisen
Harris Power Horse, 232
   owners: Donald & Alice Blom
Holt T-29, 39-41
   owner: Edith Heidrick
horses & implements, 16-17, frontispiece
   owners: Melanie & Larry Maasdam
International McCormick Super W-4, 229-231
   owner: Paul Brecheisen
John Deere Model D, 42-45
   owners: Lester, Kenny, & Harland Layher
John Deere Model B on Tractor Stilts, 218-219
John Deere Model 4850, 238
John Deere Model 8010 prototype, 236-239
John Deere Model B, 56-59
John Deere Model BWHs, 150-155
   owners: Walter, Lois, Bruce, & Judy Keller
Jumbo Simpson, 212-215
   owners: Raynard & Ruth Schmidt
Massey-Ferguson Model 98, 109-111
   owners: Donald & Alice Blom
Massey-Harris Challenger, 158-159
Massey-Harris Model 333, 181-183
Massey-Harris Twin-Power Super 101, 163-165
   owners: Wes, Bonnie & Scott Stoelk
McCormick-Deering 10-20 TracTracTor, 50-52
   owner: Edith Heidrick
Minneapolis Threshing Machine Co. Model 12-25, 37-38
   owners: Walter, Lois, Bruce, & Judy Keller
Minneapolis Moline Model U-DLX, 166-167
Minneapolis-Moline G704, 146-149
Minneapolis-Moline Jeep 210-211
Minneapolis-Moline Model GTA, 134-135
Minneapolis-Moline Model RT, 178-179
Minneapolis-Moline Model UTC, 172-173
Minneapolis-Moline YT, 202-206
   owners: Walter, Lois, Bruce, & Judy Keller
Oliver Model 88, 168-171
Oliver Super 44, 234-235
Oliver Super 66, 98-101
Oliver-Hart Parr Row Crop, 116-117
   owners: Donald & Alice Blom
Oliver Super 88, 180
   owner: Rodney Ott
Phoenix Log Hauler, 10-14
   owners: Paul & Ray Ehlinger & city of Wabeno
Plymouth Tractor, 122-125
   owner: Paul Brecheisen
Russell 20hp steamer, 18-25
   owners: Randy & Monica Sawyers
Sheppard tractor, 226-228
   owners: Ray & Dorothy Errett
Silver King Model 41, 156
Silver King R72, 154-157
   owner: Paul Brecheisen
Wallis Bear, 26-31
   owner: E.F.Schmidt
Waterloo Boy Model N, 68-73
   owners: Walter, Lois, Bruce, & Judy Keller

SUBJECT INDEX

Alvin Lombard's Log Hauler, 202, 207
Best, Dan, 19
Bosch, Robert, 85, 87
butane power, 148
C.M. Russell & Co., 18
dangers of LPG, 143-147
diesel engines, 202
Diesel, Rudolf, 78, 80, 85
Dreyfuss, Henry, 150
E.B. Wilson's Farmer's Engine, 17
Ferguson plow, 191
Ferguson System, 195
Ferguson, Harry, 187, 191, 193, 194, 195, 201
Ferguson-Sherman, Inc., 194
Firestone, Harvey, 117, 118
first gasoline tractors, 43
Ford (not Henry) tractor company, 60–64
Ford, Henry, 40, 184, 186, 188, 191, 194, 195
Ford, Paul, B., 60
gas turbine engine, 26
Guericke von, Otto, 10
Hart, Charles, 39
Holt, Ben, 19, 20, 40
hot tube system, 41
Liquefied petroleum gas, 140
Loewy, Raymond, 155, 159
LPG conversion kits, 208
Nebraska Tractor Tests, 60
Newcomen engine, 13
Nicolas Cugnot's steam road wagon, 15
Oldfield, Barney, 120
Otto, Nicolaus, 29, 30, 33
Parr, Charles, 39
Persian orange, 114
pneumatic tires, 118–123
power take-off (PTO), 207
rapid-acting steam engine, 13
Rosen, Art, 91
rubber tires, 118–123
shotgun shell ignition, 91, 207
Teague, Walter Dorwin, 155
three-point hitch, 194
Wallis, H.M., 40
Watt, James, 13, 14

431

Index

CLASSIC FARM TRACTORS

Advance-Rumely, 260–261, 267
Allis, Edward, 255–256
Allis-Chalmers tractors, 254–269
    Model 20-35, 254–255, 260–263
    Model G, 267–269
    Model M, 264–267
    Model U, 263
Allwork, II Model F, 425–427
Aultman & Taylor Machinery Company, 255, 257–261
Aultman, Cornelius, 255
Avery, Model C, 251, 415–418

B. F. Avery tractors, 379
Best, C. L. "Leo," 300–305
Best, Daniel, 249, 293–305
Best tractors, 292–309
    110hp Steamer, 294
    Model 30 "Humpback," 296
    Model 30, 301–303
Brooks, Charlie, 380
Bull Tractor Company, 371

Case, Jerome Increase, 271–273
Case Plow Works, 273–291
Case Threshing Machine Company, 273
Case tractors, 270–291
    Model 9-18, 248, 274–277
    Model 12-20 Crossmount, 278–279
    Model 25-45, 279–283
    Model LA, 283–284, 288–291
    Model VA, 270–271, 284–287
Caterpillar tractors, 292–309
    Model 15, 305–306
    Model 22, 308–309
    Model 60, 292–293, 304, 307
    Model RD-4, 307–308
Chalmers, William, 256
Cletrac tractors, 376, 378–383
    Hi-Drive Model F 9-16, 378–379, 382–383
Cockshutt, James, 251
Cockshutt tractors, 379–398
    Model 30, 398
COD, Model B, 411–413
Crozier, Wilmont, 248

Dain, Joseph, 311–312
Dain, All-Wheel-Drive, 313–314
David Brown tractors, 283
Dearborn Motor Corporation, 342
Decker, James, 255

Deere, Charles, 311
Deere, John, 311
Deere & Company tractors, 310–329
    Model 720 Hi-Crop, 328–329
    Model A, 321
    Model B, 253, 319–321, 326–327
    Model BO Lindeman, 322–323
    Model C, 250, 310–311
    Model D, 315–317, 324–325
    Model GP, 311, 318–319
    Model MC, 327–328
Deering, William, 347–349
Deutz tractors, 253, 269
Diesel, Rudolph, 250, 256
Dreyfuss, Henry, 250, 320

Emerson-Brantingham tractors, 273

Fageol, Model 9-18, 413–415
Fairbanks Morse, Model 15-30, 404–405
Falk, General Otto, 256, 261–267
Ferguson, Harry, 20, 253, 331–345, 376–377
Fiat tractors, 253
Firestone, Harvey, 250, 268
Ford, Henry, 248, 250, 331–345
Ford, Henry II, 341–345
Ford tractors, 330–345
    Fordson, 330–340
    Model 2N, 247–248, 343
    Model 8N, 343
    Model 9N, 341–342
    Model 501 Workmaster, 344–345
Froelich, John, 312

Gasoline Traction Company, 273
Graham-Bradley, General Purpose, 428–429

Happy Farmer, Model F 12-24, 406–408
Hart, Charles, 381
Hart-Parr tractors, 379–383
    Model 30-60, 380
Heider, John, 272
Heider, Model C, 272–273
Holt, Benjamin, 249, 293–305
Holt, Charles, 293

Holt, Pliny, 299
Holt tractors, 293–309
    Midget 18hp, 299–300
    Model 45, 295
    Model 75, 297
    Model 120, 298
Huber, Super Four 15-30, 418–421

International Harvester tractors, 346–369
    Farmall F-20, 253, 355–357
    Farmall Model 300, 346–347, 366–367
    Farmall Model A, 362–365
    Farmall Model H, 358–361
    Model 8-16 Kerosene, 351–352
    Model 600 Industrial, 368–369
    Model TD-14, 358
    Mogul Model 8-16, 348–351
    Titan 10-20, 352–354

Lindeman Manufacturing Company, 321
Loewy, Raymond, 250, 380

Massey tractors, 370–377
    General Purpose 4WD, 249, 370–371, 374–375
    Model 30, 376–377
McCormick, Cyrus Hall, 347–349
Melvin, C. H., 311
Merritt, Harry, 250, 268
Minneapolis Threshing Machine tractors, 257
    Model 35-70, 383–386
Minneapolis-Moline tractors, 379–401
    Model U, 399–400
    Model UTU LPG, 401
    Twin City, Model JT-O, 392
    UDLX Comfortractor, 393–394
Moline Plow tractors, 384
    Universal, 381
Monarch tractors, 264

New Holland, 253
Nichols & Shepard, 255
Nichols, John, 255

Oldfield, Barney, 250
Oliver Chilled Plow Works, 379

Oliver Farm Equipment Company, 379–397
Oliver Hart-Parr tractors
    Model 28-44, 390–392
    Model 28-50, 387–389
    Standard 70, 395–397

Parr, Charles, 381
Parrett, Dent, 371

Rock Island Plow Company, 273
Rumely, Edward, 256–257
Rumely, Joseph, 256
Rumely, Meinrad, 255, 256
Rumely, William, 256
Rumely tractors, 255–259
    OilPull Model 14-28, 257–259
    OilPull Model M, 256
Russell, "Giant" 60hp, 405

Samson tractors
    Model M, 252, 402–403, 422–424
    Sieve Grip Model S-25hp, 409–410
Satoh tractors, 252
Secor, John, 256–257
Seville, Charles, 255
Shepard, David, 255

Taylor, Henry, 255
Tenneco, 283
Terratrac tractors, 283
Toro tractors, 267
Twin City tractors, 384

Union, Sure-Grip Model D 12-25, 424
United Tractor & Equipment, 267
University of Nebraska Tractor Tests, 248

Wallis, H. M., 372
Wallis tractors, 273
    Model 20-30, 372–373
Waterloo Gasoline Engine Company, 312–313
    Waterloo Boy Model R, 312–313
White Farm Equipment Company, 379–401
White Rollin, 379
Williams, W. H., 380